"This extraordinary book offers our civilization an epochal choice between two distinct political and policy pathways to combat disastrous climate change. Huppes provides a brilliant, confidence-inspiring synthesis of political, environmental, and policy analysis."

— *Robert A. Kagan, Professor Emeritus of Political Science and Law, University of California, Berkeley, USA*

"This book gives a refreshing view on climate policy – using a true system perspective, and going back to the historical roots of the organization of current society. Gjalt Huppes, someone who I know is always looking for linking all kind of science fields, at his best."

— *Arnold Tukker, Director CML at Leiden University, Netherlands*

"This book on effective climate policy instrumentation is of immense depth of thought, written by an author of deep knowledge, with encompassing views and integrative concepts of both social and economic sciences. It comes very much in time to implement the Paris Agreement."

— *Udo E. Simonis, Professor Emeritus of Environmental Policy, Berlin Social Science Research Center (WZB), Germany*

"This book gives a delightful guide for rethinking and redesigning national and international climate policy instrumentation. Gjalt Huppes provides a persuasive political system-level analysis for calling resilient pathways to zero emission society and its governance. I believe it would be very important for both policy makers and industries."

— *Yoshinori Kobayashi, Environmental Policy Officer at Toshiba, Japan*

"Huppes' view rightly poses climate political economy in a broad and rich political sciences realm, where various institutional functions and governance settings are scrutinized. Decentralized policy making, which assumes in the real world diversified shapes, is critically proposed as the pathway to achieve effective, efficient and especially resilient low carbon pathways."

— *Massimiliano Mazzanti, Professor in Political Economy at University of Bologna and University of Ferrara, Italy*

"This book provides a deeply insightful and innovative perspective for both academic research and practice with respect to climate policy. Huppes has been contributing to the hub of diverse disciplines for overcoming the challenge of climate change"

— *Yutao Wang, Deputy Director of Fudan Tyndall Centre, Fudan University, China*

Strategic Designs for Climate Policy Instrumentation

This book provides insight into the development of effective climate policy instrumentation in two divergent and mutually exclusive directions.

Examining the role of political philosophies, the book explains why current climate policy is ineffective and unable to halt rapidly rising atmospheric concentrations of CO_2, and suggests strategies for ending the current stalemate in climate governance. Drawing on examples from real-world case studies and challenges, the author first sets out an instrumentation approach based on a command-and-control strategy that involves identifying the technologies and behaviors key to meeting the required emissions reductions, such as energy-efficient homes and zero-emission cars. The second strategy concerns institutional rearrangement, creating incentives and options that will allow for decentralized climate action. This approach would transform and strengthen current emission trading systems, such as the EU-ETS, into a price-stabilized system covering all fossil fuels, and ultimately an encompassing emission tax, as well as creating an open electricity market. These approaches highlight not only that fundamental changes in climate policy instrumentation are now vital, but also that consistent strategies such as those laid out by the author are necessary if we are to avoid costly and ineffective alternatives.

Exploring key issues such as the relationship between instrumentation and broader political philosophy, as well as applying a systems-oriented design methodology for effective instrumentation, this book will be of great relevance to scholars and policy makers with an interest in climate change and environmental politics.

Gjalt Huppes is a Senior Researcher in the Institute of Environmental Sciences at Leiden University, the Netherlands.

Routledge Studies in Environmental Policy

Why REDD will Fail
Jessica L. DeShazo, Chandra Lal Pandey and Zachary A. Smith

The European Union in International Climate Change Negotiations
Stavros Afionis

The EU, US and China Tackling Climate Change
Policies and Alliances for the Anthropocene
Sophia Kalantzakos

Environmental Policy and the Pursuit of Sustainability
Chelsea Schelly and Aparajita Banerjee

Green Keynesianism and the Global Financial Crisis
Kyla Tienhaara

Governing Shale Gas
Development, Citizen Participation and Decision Making in the US, Canada, Australia and Europe
Edited by John Whitton, Matthew Cotton, Ioan M. Charnley-Parry, Kathy Brasier

The Politics of Aquaculture
Sustainability Interdependence, Territory and Regulation in Fish Farming
Caitríona Carter

Strategic Designs for Climate Policy Instrumentation
Governance at the Crossroads
Gjalt Huppes

For more information about this series, please visit: https://www.routledge.com/Routledge-Studies-in-Environmental-Policy/book-series/RSEP

Strategic Designs for Climate Policy Instrumentation

Governance at the Crossroads

Gjalt Huppes

Routledge
Taylor & Francis Group
LONDON AND NEW YORK

First published 2019
by Routledge
2 Park Square, Milton Park, Abingdon, Oxon OX14 4RN

and by Routledge
711 Third Avenue, New York, NY 10017

Routledge is an imprint of the Taylor & Francis Group, an informa business

British Library Cataloguing-in-Publication Data
A catalogue record for this book is available from the British Library

Library of Congress Cataloging-in-Publication Data
A catalog record has been requested for this book

ISBN: 978-1-138-69630-3 (hbk)
ISBN: 978-1-315-52481-8 (ebk)

Typeset in Goudy Oldstyle Std
by Cenveo® Publisher Services

Contents

List of figures

List of tables

Acknowledgments

Books don't just appear out of thin air, but follow a path often facilitated by intensive interactions with others. My first endeavors on the subject of environmental policy instruments were in work with Bob Kagan of the University of California, Berkeley, as part of my PhD preparations, bringing focus to implementation issues. At a Law & Society meeting I attended in 1989, I first met Udo Simonis of the Wissenschafts-Zentrum Berlin (WZB) who led me back from more theoretical considerations to the practicalities of policy in a more normative context. We later wrote a first general paper on how to approach and define environmental policy instrumentation. A most stimulating long-term contact was Paul Ekins of University College London, with whom I had many discussions in the context of European Union projects about how to analyze innovation processes for policy-making purposes. Faye Duchin of Rensselaer Polytechnic and Sangwon Suh of the University of California, Santa Barbara sharpened my focus on the divide between micro-level changes and macro-level consequences, a core issue in any sustainability policy. On the same theme, Masanobu Ishikawa (Kobe University) showed the only partial connection between a formal system level of analysis and local action. Similarly, Roland Clift (University of Surrey) brought home the link between more formal sustainability analysis and the behavior of firms. Regarding the intricacies of the policy development process, Helias Udo de Haes, director of my institute, Centrum voor Milieuwetenschappen (CML, Institute of Environmental Sciences) at Leiden University, showed how compromise is necessary in the practical development process, with the ISO Standard for Life Cycle Assessment, developed in the 1990s, providing an example of how the practice may deviate from a pure and consistent base design. Sofia Simões did her PhD with me at CML on the subject of how the large numbers of overlapping policy instruments in the energy and climate domain can substantially be antagonistic: the web of instruments is to be untangled ex ante, from an integrated point of view.

The subject for this book was conceived in 2010 during informal discussions with Paul Ekins about a possible application in 2011 for climate policy and its instrumentation in the EU Seventh Framework research program. The project leadership shifted to Benjamin Görlach and resulted in a successful application for the 3-year CECILIA2050 project, which commenced in 2011. One of the tasks addressed by my institute, CML, was instrumentation design for climate policy. Discussions focused on the reasoning behind how instruments can be specified, the choice of instruments and the design of an overall instrumentation. Several project workshops addressed topics linked to many of the subjects in this book but did not yet consider the philosophical background of instrument choices. In an early CECILIA workshop at UCL in London, Marianna Mazzucato, then at SPRU–University of Sussex, forcefully advocated for the core role of government in innovation, as well as in climate policy. This subject became an important focal point in this book, but its relationship to political philosophy and governance has yet to be fully clarified.

Next, at the kind initiative of Massimiliano Mazzanti, I was invited to the University of Ferrara in 2014 as Copernicus Lecturer, where I investigated the possible roles of welfare theory in governance for developing climate policy. The Italian influence went further in a contribution to the inspiring Routes Toward Sustainability Symposium held in Brazil in 2015, which resulted in a paper on the decentralized role of cities in a book edited by Paola Spinozzi and Massimiliano Mazzanti in 2018. Cities play a more implementational role in the governance direction of Planning & Control, and are more autonomous under Institutionalist governance. I outlined for the first time the role of specific political philosophies in climate policy in a paper for the Workshop on Sustainability and Philosophy at NTNU (Norwegian University of Science and Technology) in Trondheim, Norway, to which I was kindly invited by Rita d'Oliveira Bouman, and in a paper for the Low Carbon Society Research Network conference in Warwick, England, both in 2017. More general strategic governance ideas took shape in a 2017 publication for the Dutch Wiardi Beckman Foundation (WBS), initiated and patiently supported by Menno Hurenkamp of the WBS.

My colleagues at CML contributed directly and indirectly to this book in countless projects and diverse encounters, including a week-long meeting at my house in France to work on long-term scenario development. First studies on the substance deposit as a systems-level instrument were made with Ester van der Voet and Jeroen Guinee in the 1980s, funded by Dutch and EU grants, here applied to CO_2 emissions. René Kleijn and Ruben Huele, and later Sebastiaan Deetman, have regularly given advice and commentary. After my retirement in 2011, I first worked at CML in the CECILIA2050 project and later received a grant from CML to continue the work on this book.

Of course, input also came from the home front, from countless discussions with my wife, Liesbeth Huppes-Cluysenaer, on political philosophy and governance and on the role of Aristotle in embedding political theory in broader philosophy. English corrections were first made by Jacqueline Schoonheim, with new mistakes undoubtedly added by me in a last round of content corrections, these taken care of by Donna Frassetto for Taylor and Francis. None of the people mentioned bear any responsibility for the final results and certainly not for the unavoidable idiosyncrasies having to do with multiply-linked disciplines and the substantially open domains of the book. There is no single answer to all questions, as the questions are ill defined by nature.

Gjalt Huppes

1 New instrumentation and new governance needed for effective climate policy

Introduction: climate policy instrumentation at the crossroads

Climate policy is needed

The purpose of climate policy is to counter global warming emissions. For most of the past half century, climate change has been broadly recognized as a serious, long-term challenge. It has been on national political agendas for 40 years and on the international political agenda for 30 years. Yet emissions continue to rise, with atmospheric concentrations rising ever faster since measurements were first taken at Mauna Loa, Hawaii, in 1958; see Figure 1.1. Large-scale coal mines continue to be developed, with no sign of depletion or of cost increases to come for many decades. Oil and natural gas production are also increasing due to new cost-reducing technologies, including horizontal drilling and fracking. The technical and economic conditions conducive to rising carbon dioxide (CO_2) emissions continue to exist. New developments for extracting natural gas from the ocean floor, where very substantial amounts are available, may be imminent. Exploitation will contribute not only to rising CO_2 emissions, but also to a rise in emissions of methane (CH_4), the second climate gas important to global warming. The heating effect they cause, known as "climate forcing." Industrial emissions have produced a level of CO_2 higher than any recorded for the previous million years, and are rising more rapidly than in the past *several* million years. Viewed within a geological timeframe, the world is moving into an uncharted climate domain.

Known and unknown risks

The risks of global warming dwarf other environmental risks. They concern predictions for overall warming; hotter, dryer, and wetter periods and areas; and weather instability. Dry periods with strong winds will lead to difficult-to-control wildfires. Many tropical areas will become unfit for agriculture. Melting glaciers and the expanding volume of the oceans due to a rise in temperature will ultimately lead to higher sea levels and flooded coastal areas. Rising CO_2 emissions will also inevitably lead to substantial ocean acidification. These risks are in the high probability range and are already occurring; they are known

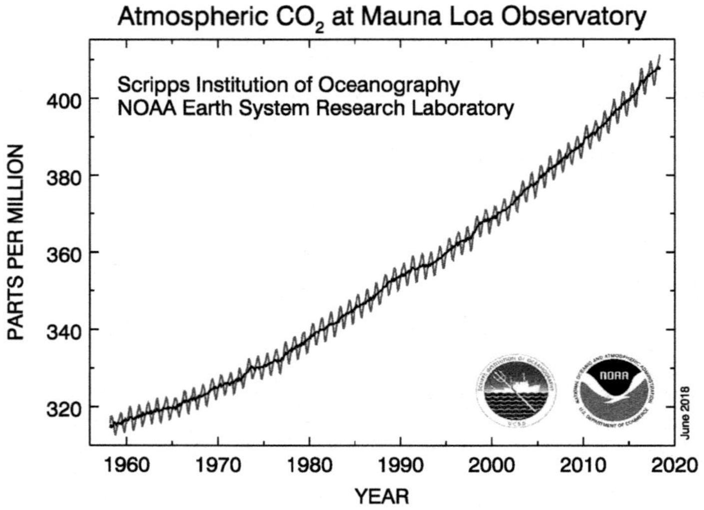

Figure 1.1 Development of CO_2 concentrations in the atmosphere, 1958–2018.

Source: NOAA[1]

and certain. In addition to these high-probability risks, more extreme effects with a lower probability loom in the future. Methane may be freed over short time periods in large quantities from tundra and ocean floors as a result of warming, possibly starting soon. This short-term explosion in emissions would be followed by an extreme rise in temperature, also short term, fueling large-scale fires in both forests and cities. A cooling period would follow, as the particles emitted in fires darkened the atmosphere. This is a runaway mechanism that has only rarely occurred in the geological record. But it has occurred in the past, and it has the potential to kill off a significant portion of the human population. This would occur through starvation, and then through the social and political disruption that would follow, including warfare. Other runaway mechanisms are conceivable, as well. These risks are incomparably larger than the risks associated with nuclear energy, including the risk of nuclear warfare. But their probabilities cannot be established with any certainty. By moving into unchartered territories, other risks, as yet unknown, may arise.

Global economic institutions for growth have reduced poverty and increased emissions

Global society has lifted billions of people out of poverty, causing a concomitant increase in the use of energy and resulting fossil fuel emissions. Over the course of two and a half centuries, economic growth fueled by industrial development was first limited to the Western world. It was only after the Second World War (WWII) that institutions began to globalize and global emissions started to

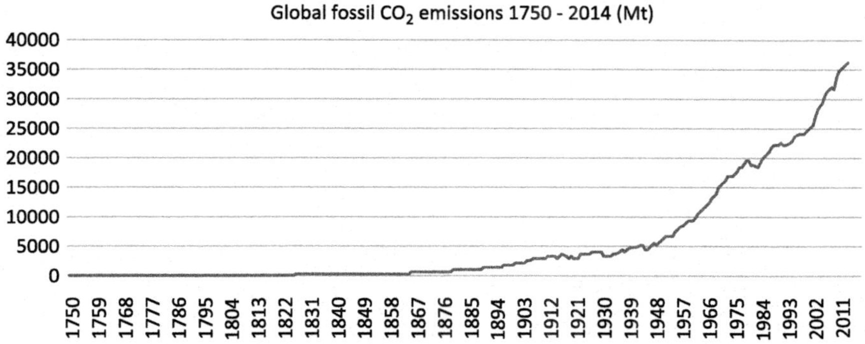

Figure 1.2 Long-term development of global CO_2 emissions.

Source of data: Boden, Marland, and Andres (2017)

explode; see Figure 1.2. Growth first expanded to countries in the Western alliance: Japan, Singapore and Taiwan, followed by Korea. Global growth in recent decades has been dominated by non-Western countries, with India likely following in the growth path of China. The technologies, knowledge and institutions for continued economic growth exist. All developed and developing countries have rules and organizations for education, basic research, research and development (R&D) and investment. Global institutions needed for international trade and investment and for knowledge exchange have been established, as well. Economic growth could theoretically eradicate extreme poverty across the globe in a matter of decades. However, this desirable but unprecedented economic development must be accompanied by extreme emissions reduction, which is also unprecedented and as yet sorely lacking.

Technologies for achieving zero-emission societies already exist

The technical means for creating low-emission societies are already largely available. The technical means for reducing emissions by 90% over the next 50 years exist today and can be further developed both technically and economically in the future. Economically speaking, the cost of non-fossil energy is now in a range comparable to the cost of fossil energy historically. Wind and solar heat and electricity production could take over completely from fossil energy use, with other energy technologies such as wave, hydro, nuclear and geothermal possibly developing, as well. Wind and solar energy production have expanded rapidly but as yet scarcely contribute to total primary energy production. Even were these options to be expanded more rapidly, a basic problem would remain: while these energy sources add to total primary energy production, they only partly replace fossil energy use. Fossil energy remains abundantly and cheaply available even in the long term. Carbon capture and storage (CCS) technology could reduce fossil emissions by permanently storing CO_2 underground. However, this option requires costly installations for CO_2 capture, transport and storage, and requires substantial amounts of energy, as well. Where CCS is possible on a large scale, as in fossil electricity production, the costs of electricity would rise by one third at least, with roughly a quarter of the emissions still remaining. Direct air capture and storage (DACS) would have similar costs. But which country would be willing to pay the pure costs of CCS?

Environmental policy instruments have been effective – but not for climate policy

Policy instruments to solve environmental problems were developed in the 1970s and 1980s. Instruments include operating permits, technical standards, emission standards, tradable emission permits, taxes and subsidies, technology development, and better information on clean options. Where political pressure was great enough, these instruments have been employed effectively. In industrialized countries the main environmental problems related to acidification, toxicity, and over-nitrification have thus been substantially resolved, often with emission reductions of well above 90%. The earth's protective ozone layer also began recovering once main ozone-depleting emissions were restricted under the Montreal Protocol in 1987 (with recently some emissions rising however, from as yet unknown sources). If these successes have been possible, why is resolution of the climate problem of global warming so elusive? The need for effective policy is clear, and this need will not diminish anytime soon. The technologies exist and can be further developed. The policy instruments also exist or can be developed. Why, unlike other environmental problems, has the climate problem not been resolved?

The climate problem is a "wicked problem"

The climate problem is unique. It is a particularly thorny environmental problem that has been described as a "wicked problem" or even a "super-wicked problem" (Levin, Cashore, Bernstein, & Auld, 2012). A wicked problem is defined as a problem for which no clear single solution is available (Buchanan, 1992; Rittel & Webber, 1973). The climate problem is a wicked one due to the different natures of the long-term risks, the fact that measures are effective only over the long term, and the global nature of the problem with respect to both its causes and effects. Virtually all economic activities – and certainly all countries – contribute to global warming. These factors contribute to the difficulty of choosing a single effective solution. Well-founded disagreement on the best solutions or even approach easily leads to postponing adoption of any effective policy measure. One single instrument type, such as CO_2 emission pricing for example, can be implemented via mutually exclusive instrument options – tradable emission permits versus emission taxes. And even more difficult, the solution lies not in one or a few single instruments, only. A set of mutually supportive instruments is required, to be implemented continuously over the course of the next half century, with substantial revisions of current non-climate regulations and infrastructure development.

International alignment on instrumentation is also needed, including effective sanctions against countries that reap undue competitive advantage from their failure to implement climate policy. This international component further adds to the wickedness of the problem as there is no global government to effectively enforce global rules. The United Nations (UN) has no executive apparatus to implement and enforce regulations: nation states are pretty much independent.

Growing inequality creates additional political constraints in Western countries

The political constraints caused by growing income and wealth inequality in recent decades add to political wickedness. Those left behind hardly have a reason to support global climate policy. This broader sociopolitical problem must somehow be resolved. Placing the climate policy task into a longer term historical context adds further restrictions and complexity to possible solutions, increasing the political contribution to wickedness. A substantial body of historical empirical research has examined how societies open up politically to broader segments of the population, allowing them to flourish and grow, but then close off as economic concentration of wealth leads to political power of the richest (Acemoglu & Robinson, 2012; Piketty, 2014; Van Bavel, 2016; Scheidel, 2017). Primary income inequality is on the increase in most countries of the world, with wealth increasingly concentrated in the hands of a tiny fraction of the richest 1%, the super-billionaires. In these countries, safeguarding developed positions becomes a primary concern in steering politics, leading to further monopolization. In democratic countries, the lower income half of the population now tends to lose out, with their relative and even absolute primary income positions deteriorating. Thus, a substantial part of the electorate is dealing with issues other than climate policy, while the rich see their primary task as safeguarding their individual positions in a competitive world. The lack of legitimacy resulting from increasing inequality is present in less democratic countries, as well. Without broad legitimacy, climate policies which by necessity conflict with specific interests are difficult to develop and implement. And if implemented, they may well break down in adverse times. It would seem that solving the climate problem will require deep political changes.

New instrumentation design can take one of two different directions

Depending on whether one or the other basic choice of governance direction is followed, different instrument sets will come into play. Focusing on institutions links generally to an institutional redesign, whereby combined economic and political power concentration is reduced and then instruments to create general incentives for emission reduction are developed. The move is away from increasingly centralized government with increasingly dense regulations. This approach aims for generic market revision first, to resolve the market failures leading to and exacerbating the climate problem. The other approach takes the new technologies required for emission reduction as a starting point and envisions how old technologies can be phased out and new ones phased in. The reasoning behind the second, more planning-oriented approach is based on broadly accepted welfare considerations: doing what is best for society at large, and for a maximum number of people and organizations. In this approach, victims – those who will lose out as a result of policies – can be compensated from the overall welfare gain. Denser regulation is not a goal, and is to be avoided as much as possible, but must be accepted if needed. The two approaches have barely any overlap with respect to instrumentation. A deep choice is necessary to ensure consistent development. We are standing at a crossroads regarding the development of effective climate policy instrumentation.

Effective instruments in context: socioeconomic and political options and constraints

New instrumentation must be placed in the relevant socioeconomic and political context, including the developing inequalities in wealth and power, globalization, and information technology (IT). Which instrument strategies may fit within the particular economic and political landscape at hand? This fundamental question is multifaceted and can be answered on various levels: first, at the level of political philosophy; next, at the level of corresponding governance approaches; then by looking at the link to the socioeconomic and the political reality as it is unfolding; and finally, at the level of specific options for instruments and overall effective instrumentation.

One option is to develop a view of governance that reverses today's growing tendency toward oligarchy and breaks up some of the dominant vested interests – similar to the process that occurred at the end of the 19th century in the United States, when the "robber barons" were forced to break up their monopolies. New coalitions of traditionally opposed political views will be required if this is to occur. This option is linked to a political philosophy in which value is placed on avoiding centralized power and monopolies. The other governance option is to stick to current piecemeal policy development, overcoming specific oppositions one by one within the currently dominant sociopolitical structure. This option is linked to a political philosophy in which the state does what is best from the perspective of the broader welfare, in both social and climate policy. Income effects of climate policy are reckoned with in instrumentation design. The challenge of how to develop well-reasoned instrumentations for what are termed the *Institutionalist path* and the *Planning & Control path* is the subject of this book.

Why new climate policy instrumentation is needed

The basic reason for developing new climate policy instrumentation is that the old instruments have failed to deliver meaningful emissions reduction. Atmospheric concentrations of CO_2 are rising more rapidly than ever. The backdrop to this rise is the explosion of global economic growth, based on institutions that hardly anyone is willing to sacrifice. The instruments for emissions reduction that have been developed, from feed-in tariffs to emissions standards, have been effective each in their own micro-level domain but have not delivered overall reductions. National and global market mechanisms led to a rebound, the net effect being dominated by economic growth. The basic ideology and instrument mechanisms that substantially resolved earlier environmental problems do not effectively apply to the climate change problem in the new socioeconomic and political structure that is developing. A rethink is required for policy design and instrumentation, starting at the level of political philosophy and governance, adjusting to the new structure and guiding its further development. New policy for the global warming problem must overcome the current lock-ins to the well-developed partial and incidental instruments of old.

Old instruments were effective at a micro level but not at a macro level

Current climate policy instruments have not delivered, but not because they did not work. Feed-in tariffs on solar photovoltaic (PV) cells did, in fact, stimulate development of solar PV substantially, and wind power benefitted through similar subsidy instruments; fleet standards for personal automobiles succeeded in reducing emissions of the fleet sold each year, at least in terms of the standards; and coal-fired power stations that were closed no longer emit CO_2. These results are impressive in a technical and micro-economic sense. Large-scale production of zero-emission technologies induces technical improvements and reduces costs dramatically. Solar PV cells were roughly halved in price with every fourfold increase in production. The International Energy Agency (IEA) and the International Renewable Energy Agency (IRENA) have surveyed technical developments both broadly and in detail in numerous publications.[2] A discrepancy remains, however, between micro-level improvement and macro-level performance, mainly due to difficult-to-control economic mechanisms and broader rebound mechanisms, as well as contradictory regulatory actions. Solar electricity does not replace fossil energy, but adds to primary energy production, crowding out not only some fossil fuels, but also some clean hydro and wind power. The wind does not always blow and the sun does not always shine in large areas; thus, fossil fuel power stations that operate at a loss are now often subsidized as backup facilities. This subsidy takes the form of a *capacity payment* to prevent them from closing down. They are used when advantageous based on short-term marginal costs, emitting CO_2.

Emission pricing has developed in the European Union Emissions Trading Scheme (EU-ETS), the globally most advanced pure cap-and-trade system, whereby a capped amount of emission permits can be traded for use by the emitting firms involved. Subsidies for renewables reduce emissions by decreasing other electricity use, partially fossil based, as does closing down of old coal-fired power stations. But emissions under the cap-and-trade system will then not be reduced. The permit price will go down, incentivizing others to emit more, up to the level of the effective cap. These instruments are contradictory. The EU originally proposed a supranational emission tax (Delbeke, 1992) that would not have this undesirable effect. However, the proposal ran into political problems. Supranational taxes were a "no-go" for several countries (Andersen & Ekins, 2009), including the United States – a country with which it is preferable to be aligned when it comes to climate policy instrumentation. Other technical measures have been made less effective than policy efforts show. For instance, cars in the EU have been "optimized" for test drives, as defined in the fleet standards, with actual performance deviating substantially.

Broader rebound mechanisms also exist. Consumers in the United States now prefer less-regulated small trucks over personal automobiles that are regulated by stringent fleet standards. The closing of coal-fired power stations is often followed by the construction of new production facilities, wherein choice basically depends on the cost ratios for different fuel systems. If coal is relatively cheap it will supplant natural gas in the market, as happened in highly developed countries like the Netherlands when the lower price of U.S. coal led to reduced coal prices in Europe. Overall, coal remains attractive for electricity production due to its stable low prices and nearly endless supply, while oil and gas production are becoming cheaper every decade. Emissions pricing would tend to eliminate coal first and natural gas later, once the price had risen sufficiently.

Regulating emissions directly is somewhat like squeezing a rubber ball or balloon: only when pushing against the ball on all sides at the same time can the ball be compressed. In an administrative-technical sense this is a primary option, but it requires an integrated global view of encompassing instrumentation – a systems view that must be translated to all emitters at the national level. This systems view is one major line of new instrumentation investigated here, based on better planning and control. The main obstacle to its success is the significant amount of direct regulation required. Increased regulation is not only opposed by the regulated entities involved, but also runs counter to dominant political views in democratic countries, supported by a large share of the voting population. Existing instruments have not failed because they did not work. They failed because they were not applied broadly and stringently enough, as is required for solving the climate change problem. The causes of macro-level failure might be overcome by mending these defects with new instrumentation design. To do so effectively, however, resistance must be overcome to new, detailed, and encompassing regulatory instruments through better – *simpler* – design, as advocated by Cass Sunstein (2014).

The other instrumentation option is to redesign the overall incentive structure in society. This redesign must also be done in an encompassing way to cover all economic activities with direct CO_2 emissions and, indirectly, all others through market mechanisms that, for the most part, already exist. A step in this direction was taken by setting up trade in CO_2 emission permits, intended to establish a general price for most emissions. The EU-ETS is the primary example. It has more or less collapsed due to irrelevantly low prices, covering only a part of the emissions in the EU, with substantial administrative burdens, and continues to be linked to fraud (Interpol, 2013). The low price has several causes:

an oversupply of permits due to political vulnerability; the effectiveness of other policy instruments, especially subsidized renewables in markets covered by the ETS or linked to it; lower economic growth than expected; and shifting of material production, and its emissions, abroad. A more detailed discussion of this issue follows in Chapter 3.

The two directions for instrumentation investigated here are, first, a leaner but more encompassing option for effective regulatory measures to make them feasible and acceptable, and, second, novel options for broadly applicable redesign of institutions for deep emission reduction. Each of these directions for instrumentation aligns with a different view of how more general governance must develop, based on differing political philosophies. These views have been present in political theory since the time of Plato and Aristotle, as discussed in Chapter 2.

Atmospheric CO_2 concentrations are rising ever more rapidly

Climate-changing emissions have been the result of economic growth, which has been rising since the Industrial Revolution as energy requirements were initially met with fossil fuel–based energy and, later, through the addition of hydropower and nuclear energy. Global economic growth began to accelerate after the Second World War – with large growth spurts in the Organization for Economic Cooperation and Development (OECD) countries from the 1950s and in East Asia from the 1990s – becoming fully global after China entered the World Trade Organization (WTO) in 2001. The establishment of new global institutions spurred economic growth: a well-organized national and international monetary system, liberalized international trade and rapid technological development supported by publicly funded basic and applied research. The increase in global gross domestic product (GDP) averaged more than 4% in the half century from 1965 to 2015[3] somewhat higher than the rise in the primary energy supply. Fossil CO_2 emissions rose more or less in line as well, at around 3% per year.[4] However, cumulative CO_2 emissions in the mere 20 years from 1995 to 2015 were higher than those in the 225 years prior, from 1750 to 1995 – that is, since the start of the Industrial Revolution and large-scale coal mining; see Figure 1.2. The rapid rise of CO_2 concentrations in the last half century is unprecedented in the geological record with perhaps one exception: the Paleocene-Eocene Thermal Maximum (PETM) 56 million years ago, which may have been caused by a large comet strike (Schaller, Fung, Wright, Katz, & Kent, 2016). The next ice age, which had been advanced as a possible menace 50 years ago also due to the cooling effect of sulfur emissions, is no longer approaching. Instead, a rise in both land and ocean temperatures is continuing; see Figure 1.3. CO_2 concentrations during the last ten ice ages (800,000 years) have varied between 180 and 280 parts per million (ppm) (Pages, 2016). Current concentrations are more than 400 ppm and rising rapidly, bringing the earth into unchartered and irreversible domain, as emitted CO_2 remains in the atmosphere for centuries. The resulting rise in temperature may, in the short term, begin to free biotic CO_2 from forests fires and methane from melting permafrost, possibly creating a runaway effect in global temperature rise.

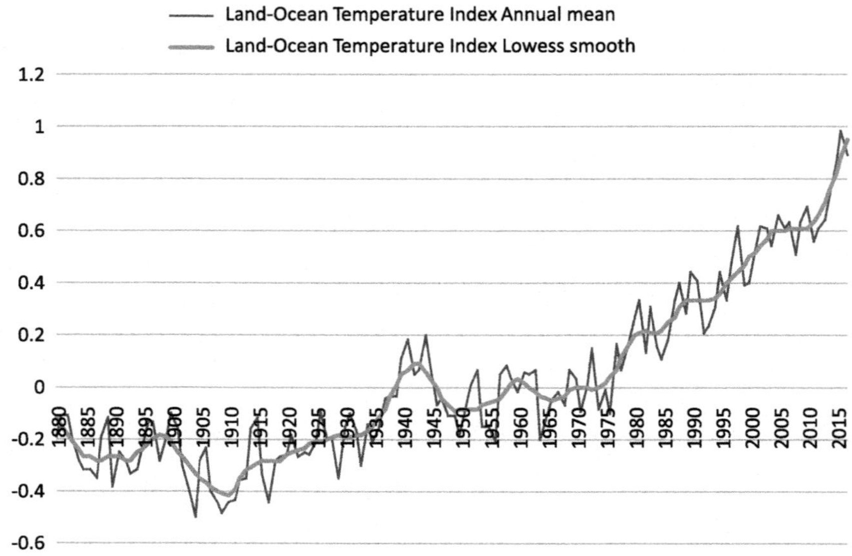

Figure 1.3 Long-term development of global temperature changes.

Source: NASA/NOAA[5]

Note
Land–ocean temperature index, with base period 1951–1980. The dark grey line is the global annual mean and the light grey line is the 5-year Lowess smooth.

The scientific and political agenda has been adequately developed

The basics of climate change are well known and undisputed. CO_2 is a greenhouse gas, as Arrhenius and Holden (1897) convincingly proved, including by atmospheric measurements. CO_2 molecules do not influence the inflow of high-frequency solar radiation but instead restrict the outflow of infra-red radiation from the earth. Incoming high-frequency solar radiation is transformed into heat with low-frequency infra-red radiation, resulting in radiative forcing, the greenhouse effect. The increased difference between energy coming in and energy going out from the earth unavoidably raises the temperature of the earth system, including the atmosphere and oceans. The basics of the unfolding climate problem were already clear in the mid-1970s. Flohn's survey paper in *Climatic Change* indicated quantitatively that a very serious climate problem in the following decades was in the making (Flohn, 1977). Data availability on emissions has greatly improved since then. CO_2 is the major greenhouse gas in terms of quantity and also by effect, accounting for around two thirds of emissions (Intergovernmental Panel on Climate Change [IPCC], 2014, p. 7). Methane contributes a substantial share now, with nitrous oxide (N_2O) following and both are increasing. Much more detail in climate modeling has been added since then as well. Major improvements were the result of links to adjoining mechanisms, such as global cooling due to sulfur dioxide emissions, the role of other albedo changes, the role of ice melting and mixing mechanisms in the oceans, and feedback mechanisms between economy and climate.

The IPCC has surveyed and helped standardize scenario modeling to make modeling outcomes comparable. The IPCC is the global forum for scientists in the climate modeling domain and is supported by national governments, science foundations and universities. Projections indicate that a passive business-as-usual scenario without further climate policy will lead to extreme temperature rises on the order of 8 degrees Celsius by the end of the century. Eight degrees is a median value, with a probability of not exceeding that value usually set at 65%. The risks of runaway effects leading to even higher temperatures are present but difficult to quantify. These risks certainly increase with a higher and faster rise of the average global temperature. The freeing of methane from tundra and ocean clathrates could form mechanisms for such runaway effects and lead to even more extreme short-term temperature increases. Large-scale burning of drying tropical forests may also contribute to runaway effects. Task-setting scenarios keep temperature rise below a chosen threshold with a certain probability, to be politically decided. Currently, the most generally agreed target-threshold is a maximum temperature rise of 2 degrees Celsius relative to preindustrial levels. To further reduce risks, 1.5 degrees as the median probability rise is preferable, as set forth in the 2015 Paris Agreement advocating "well below two degrees." The path to achieving this limited rise must be politically acceptable and economically feasible. Scenarios for a 2-degree future, referred to as *RCP2.6* scenarios in the jargon of the IPCC, therefore tend to start too slowly, first overshooting the 2-degree mark, with negative emissions to be realized in the second half of the century (IPCC-AR5-WGIII, 2014, Figure 6.7). The 1.5-degree goal may already be unattainable because of the risks of economic and policy breakdown; therefore, some see it as wiser to work toward the already challenging 2-degree goal (Jordan et al., 2013). However, negative emissions in the 2-degree scenarios are based, for the most part, on the assumption that large amounts of low-emission biomass would be incinerated using CO_2 capture and storage of the biotic CO_2, an assumption whose viability is clearly questionable (Heck, Gerten, Lucht, & Popp, 2018).

Runaway effects with short-term extreme temperature rises would lead to extreme food shortages and the breakdown of many technical systems, followed by a breakdown of political systems, with epidemics raging across the globe: the biblical Four Horsemen of the Apocalypse at large (Scheidel, 2017). This is the *fat tail* of the cost of catastrophic climate change (see Weitzman, 2009, 2011): a low chance but high damage effect, not susceptible to quantification in economic terms. Such disruptive effects may be of overriding importance from a policy perspective, dwarfing any other calamity for which society is responsible – not comparable to large-scale nuclear reactor meltdowns or massive earthquakes destroying hydroelectric dams, or even global nuclear war. But as with other catastrophes, they are difficult to model predictively and cannot be specified in economic terms for that reason, as well. With higher CO_2 emissions, not only will the occurrence of "normal" damage rise, but also the probability of low-chance, high-impact effects. That outcome is certain, although not quantifiable.

The current rise in CO_2 concentrations sets the world on a path to severe damage and to the possibility of an apocalyptic catastrophe unlike any that has occurred in our often deeply tragic human history.

International developments in climate policy are not effective

Global warming became more visible in the 1980s, as global land surface warming overcame the sulfur-related aerosol cooling effect of OECD economic growth, retreating glaciers and a measurable rise in sea levels. The IPCC was set up in 1988 to gather scientific information regarding climate change. Supported by more detailed modeling, international policy efforts began. The OECD translated its extended "polluter pays principle" into practical options for climate policy instrumentation (OECD, 1992). The United Nations Framework Convention on Climate Change (UNFCCC, 1992, [https://unfccc.int]) focused at first on developing and transmitting better information about emissions. It laid out the route for international policy coordination in regular Conference of the Parties (COP) meetings. This process led, in 1997, to ratification of the Kyoto Protocol, which obligated most developed countries to reduce their emissions. However, the United States, the largest emitter at the time, did not ratify the agreement. Raising objections, the U.S. government required commitments for scheduled emission reduction from major developing countries and refused to accept serious harm to the U.S. economy, the largest energy producer in the world (Hovi, Sprinz, & Bang, 2012).

In the years since, China has become the largest emitter, making it much less useful to enter into binding agreements that do not include both the United States and China. More recently, the Paris Agreement (2015) reached at COP21, replaced the binding national emissions reductions of the Kyoto Protocol with national pledges. However, the pledges on emission reduction are not enough to restrict global temperature rise to well below 2 degrees Celsius, as recommended in the Paris Agreement (the 2-degree or 1.5-degree stabilization goal). Subsequently, the United States has retreated from its pledges, and Germany has been missing its targets. Although the IPCC has coordinated scenarios for emission reduction over the past two decades, these scenarios tend to technical and economic analysis, and do not flesh out the required policy instrumentation in any detail, let alone distinguish divergent views on governance in instrumentation design. The question of new instrumentation design for single countries and internationally has not yet been addressed.

National and international institutions generate economic growth and emissions

The core mechanism of economic growth is higher efficiency – utilizing lower amounts of inputs per unit of output, thereby making more with less. Inputs are ultimately factor inputs: labor and natural resources together with the capital goods produced by labor and natural resources. Population growth will add to labor. Capital goods and IT also create better access to natural resources. Efficiency increases are created by innovations in technology, improvements in management, better health, better education and the creation of general knowledge (Mokyr, 2004). Markets actively developed and supported by an institutional and legal system play a key role in the overall alignment of economic activities. These factors leading to increased efficiency are embedded in institutional arrangements that are publicly organized and supplied, including universities and research institutes, the educational system, health care supply and infrastructure management – all dependent on the most general of institutions: a country's constitution. It forms the backbone of society, establishing property rights, conflict resolution, public decision making and the options and incentives for adapting to new developments. The infrastructure necessary for the transport of people and products is largely publicly supplied or regulated, as for example via zoning laws and roads and ports. Indirect or roundabout production, as capital formation, contributes in a major way to increased efficiency. The production of bread is based on the wheat from which it is made, for example. But virtually all economic sectors of all countries in the world have delivered indirect input for producing the bread: from mining and steel furnaces to power plants and lubricants, from transport infrastructure and what floats and rolls through it, from software to control processes, and so on.

Energy efficiency increases achieved through the use of steam pumps started the Industrial Revolution. Many documents created by national governments as well as the IEA and OECD advocate accelerating increases in energy efficiency as a way to reduce emissions. The micro-reasoning for increasing efficiency, using less input for a given output, is mirrored at the macro level; however, by having more economic output for given factor inputs is how economic growth is defined. The basic mechanism of efficiency improvement is the basic mechanism of economic growth. Historically improvement of energy efficiency has led to more energy use, not less, with some relative decoupling of the two only since the 1980s, but no absolute decoupling. Energy use still rises with economic growth, and fossil energy is still fully dominant in the growth of the total primary energy supply. This is the macro-level reality that must be changed.

The Jevons paradox, presented by economist William Stanley Jevons in 1865 in his book *The Coal Question* (Sorrell, 2009), links the micro and macro levels: the dramatic increases in energy efficiency in the 18th century did not lead to reduced energy use. Instead, they were the drivers of dramatic increases in energy use, triggering the climate problem caused by mining and burning coal on an ever-larger scale. The relationship between economic growth, growth in energy use, and growth in emissions is not completely rigid. Micro-efficiency improvements can outpace economic growth, or not. Cars can be made more fuel efficient. But doing so also makes them cheaper to drive. Some of this increased efficiency is thus spent on increased driving in such fuel-efficient cars. The other part is a broader income effect; the income generated by increased efficiency will be spent on something else. Air travel has a high income elasticity of demand, meaning that when people's incomes rise, they tend to spend a higher share of their income on air travel. This is currently the case in both developing and developed countries for all levels of income (Goodwin, 2012). Effective fleet standards for cars may thus help increase aviation. The main institutions for efficiency improvement and economic growth will certainly remain in place: the legal system, public education and research, the financial system, and the international trade system. It is the more general institutions that can be changed, or the highly detailed and stringent regulations that are to come. Without such changes, and barring economic and political disasters, CO_2 emissions will continue to rise over the course of the next half century just as they did in the last half century. New, far-reaching climate policy is required to prevent impending climate disaster, with new instrumentation to override these basic mechanisms for economic growth with emissions.

Why did global economic growth explode after WWII? The answer lies in features of the general institutional framework in place in the industrializing world: contract law, liability law and trade law, broad public provision of education, publicly funded research, basic infrastructure, social security, a system of health care, a well-developed tax system, the rule of law, and much more. All industrialized countries in the OECD share these characteristics. More specific causes can be found in events occurring at specific points in time. The Marshall Plan to rebuild Western Europe (1947) helped national economies recover from the destruction of WWII with monetary support and support in kind. New national monetary systems were established, leaving behind the rigidity of the old gold-based system; substantially independent national Central Banks were installed with the task of stabilizing currency and controlling the banking system; fiscal policy and public expenditure began to regulate the business cycle based on new macro-economic insights and modeling; and these innovations were also used to stabilize the current account of the balance of payments of Western countries. The International Monetary Fund (IMF, 1944) created a relatively stable international monetary system suited for economic growth, adapted in due time to new tasks. The World Bank (1944) began to fund otherwise difficult to finance basic investments, especially in large-scale infrastructure. The General Agreement on Tariffs and Trade (GATT, 1947) opened up international trade, countering policies that led to collapse in the 1930s when defensive trade restrictions were imposed by the United States.[6] Tariffs and non-tariff restrictions were reduced, with a major reduction occurring in the Kennedy Round in the 1960s. The GATT led to the development of the WTO (1994). The United Nations (1945) became the general forum for international cooperation and safety. The more developed nations set up the OECD (1960) as a follow-up to the Marshall Plan. These new international institutions were the basis of recovery and growth after the global depression of the 1930s and the destruction of the WWII. China entered the WTO in 2001, accelerating globalization and global economic growth once again. India is moving in a rapid-growth direction as well, as are several other formerly underdeveloped countries, especially in Southeast Asia. Between 1960 and 1980 economic growth in developed countries entered the double digits, with very rapidly expanding oil and coal production. From the late 1990s onward, broader globalization led to double-digit growth in China, which in turn led to rapidly expanding production of coal, oil and natural gas globally. International trade as a percentage of global GDP is now back to the same level as it was in 1929 when the stock market collapsed, throwing the developed world into a major economic depression, and international trade was halted also by the United States Smoot-Hawley Tariff Act of 1930.

The unprecedented economic growth of the past half century also led to an unprecedented rise of CO_2 emissions and atmospheric concentrations of CO_2. From an emission perspective, the global industrial revolution started not in 1750 but in 1950. Climate policy instrumentation must be placed in this global context of development toward sustained growth, which is essentially a new task. Halting growth is not an option as long as poverty remains, and would certainly not lead to a deep reduction in emissions. It is also not a popular route politically, because it could easily lead to an erosion of stock prices, tumbling housing prices, large-scale bankruptcies, and explosive poverty. There are good reasons to reduce growth in rich countries, but they are not essentially related to climate change. Tim Jackson points to negative welfare effects of economic growth (Jackson, 2011). The "Degrowth" movement advocates more basic, deep institutional changes away from internationalism, liberal capitalism and growth (Kallis, 2011).

The climate problem was first recognized as an environmental issue related to global economic growth around 1970, in parallel with toxic emissions to water, air and soil; eutrophication; and the effects of changing land use. Land use change, at first primarily viewed as a problem involving loss of the natural world, also contributed substantially to CO_2 emissions. These emissions are no longer rising now and may already be declining somewhat, but they still contribute to rising CO_2 concentrations. The problem of toxic emissions having a regional to continental range, including massive emissions of sulfur dioxide and nitrogen dioxide, has been substantially resolved in developed countries. Ozone layer–depleting substances have effectively been phased out globally. This is not the case with fossil CO_2 emissions. Although the Kyoto Protocol was a good start for global climate action and the Paris Agreement reflects the good intentions of many countries, intentions do not constitute implementable policy, let alone implemented policy, not to mention effective policy. Emissions continue to rise, leveling somewhat only in times of recession, and atmospheric CO_2 concentrations continue to rise relentlessly, as Figures 1.1 and 1.2 show. A rise in air temperature follows the rise in CO_2 concentration, with a time lag of about a decade. Surface air temperatures rise faster than the water temperature of the oceans, and the rise in surface air temperature is more rapid in the arctic than in the tropics. It is the global climate system that is warming, with rising global concentrations of greenhouse gases forming the core causal mechanism, and fossil CO_2 being the dominant greenhouse gas.

So why are new designs for climate policy instrumentation needed? The short answer is that current instruments have not delivered the emissions reductions of CO_2, methane (CH_4) and nitrous oxide (N_2O) that are required for climate stability. Other emissions-related environmental problems have substantially been resolved in the last half century, including large-scale emissions of sulfur (SOx) and nitrogen (NOx). We now have a half century of experience in developing and implementing environment and climate policy. However, the instruments used do not appear capable of doing the job of tackling global warming. To understand why, we need to answer a number of questions: why have the instruments failed to deliver on climate policy? What are the causes for the continued rise in emissions and atmospheric concentrations of CO_2 and other greenhouse gases? Can these causes be managed? What technical options are there for emissions reduction and how might they be realized? What economic constraints hamper effective policy, and how can they be overcome? And, ultimately, what are possible social and political constraints to setting up adequate climate policy instrumentation, and what strategies could be designed to effectively overcome these constraints?

But first, we need to ask: why must emissions be reduced so quickly? Are matters really so urgent now? For action to become politically viable, acceptance of the urgency of reduction is imperative. Urgency has increased in recent decades as knowledge about climate change and its risks has developed. The very long average atmospheric life time of CO_2 means that current emissions will continue to exert their climate-forcing influence for centuries, irreversibly. This is less so with the second most short lived climate gas, methane. The third important climate gas, nitrous oxide, also has a long lifespan in the atmosphere, even longer than that of CO_2. The link between climate forcing and temperature rise has now been established more definitely, and in more quantified detail, taking into consideration other mechanisms such as the global cooling effects of sulfur emissions and albedo changes. Data on past CO_2 emissions and concentrations, and climate change in the geological record, have also become available, including instances of very rapid temperature rises. Insight into how deep the emissions reductions must be to avoid unacceptably high risks has increased greatly, based on better modeling and a reassessment of the risks of low-probability, high-impact runaway effects of climate change. There is broad political consensus around the goal of limiting the rise in temperature this century to well below 2 degrees Celsius, and preferably below 1.5 degrees. To achieve this goal, the total emissions budget for the remainder of this century cannot exceed 800 gigatonnes (Gt) of CO_2, regardless of a specific emissions time frame, since any unit of CO_2 will exert its climate-forcing effects for centuries. The 2-degree/1.5-degree goal is a political choice. It does not represent a fixed prediction but is a central value in a much broader probability distribution. Continuing with current emissions on the order of 40 Gt per year, the budget would be spent in 20 years' time. Translating the budget into a policy task, emission reduction of around 90% would have to be achieved in the next few decades. Overshooting this budget mark already seems inevitable. The realistic design of viable instrumentation has lagged, with atmospheric CO_2 concentrations rising faster than ever. Without new instrumentation this extremely difficult task is hardly feasible. How to develop such instrumentation as a feasible option is the task set for this book. Two basically different and substantially incompatible options for instrumentation are distinguished.

Globalized institutions have led to an unprecedented rise in global emissions

Cumulative emissions in the 25 years between 1950 and 1975 were higher than the sum total of emissions produced in the previous 200 years. Figure 1.2 charts these long-term trends. Emissions rose somewhat more slowly in the two decades after 1975, but continued to increase steadily. They accelerated again with the spurt in global economic growth, especially in China, as ever-faster growth of international trade brought industrialization to many countries around the world. Poor countries also have begun to develop economically, with only a few lagging behind. In the quarter century from 1990 to 2015 cumulative emissions were again higher than the sum total of all emissions from 1750 to 1990. If these trends continue apace, maintaining a business-as-usual approach, the total emissions from 2015 to 2050 will again dwarf the sum total of all emissions from 1750 to 2015, bringing atmospheric levels from 280 ppm to well over 500 ppm. The causes of economic growth may be centuries old, but what is new are the volume of growth and the size of concomitant emissions and environmental problems. This change in order of magnitude is what has taken billions of people out of poverty, and it is also what has made greenhouse gas emissions a global problem. No one would want to deny development to those still living in poverty. The relatively poor in all countries strive for higher incomes, as do those with already higher incomes. But without effective climate policy, successful economic and social policy will again lead to increased emissions roughly in line with rising incomes and will be reduced by the decoupling of energy use from growth in GDP. This reduction has occurred in the last three decades, but not in the three centuries before, and could very well be reversed due to the rapidly diminishing cost of fossil energy production as well as by fracking and open pit coal mining.

Resolving the clash of political philosophies in climate policy instrumentation

Current instrumentation hinges on the hybrid mix of two opposed political philosophies, as illustrated by cap-and-trade systems such as the EU-ETS. It includes planning and control in that each firm has to have emission permits for its emissions, the total of emissions is fixed and control is based on emission measurement, using a fossil fuel administration for each firm. However, the substantial burden of administration makes the instrument unfit for smaller firms. Institutionalism is involved, since a generic price evolves creating an incentive for cost-effective emission reduction, also longer term. However, the inelastic supply of permits (their volume is fixed in advance), combined with their short-term inelastic demand, would create price volatility, with alternating periods of very low and very high prices. This makes long-term price expectations opaque. The planning and control entry to instrumentation does not stop at a single instrument, of course. Added are emissions standards, such as those for power stations, and various subsidies for renewable electricity. These instruments, effective apiece, reduce emissions under the cap, but not the cap itself. So they do not reduce emissions but only reduce the price of the emission permits. There may be administrative technical solutions to this problem, such as adapting the cap to raise the price. But the result is more complexity in administration, with substantial political requirements, as the EU is experiencing. The ineffective instrument mix based on substantial administrative and political efforts seems a dead-end alley, to be exited as soon as possible. A clear starting point in political philosophy and adjoining governance directions seems a minimum requirement for success.

This conclusion holds unless autonomous developments are able to reverse the rising trend. This would require not just that net emissions level off, but that they drop over the course of the next half century to near zero to reach the 2-degree climate goal, and even faster to meet the 1.5-degree goal. Is it possible that this reversal could happen autonomously, without active public policy?

Could CO_2 emissions cease autonomously?

Various optimistic arguments can be made for why the climate problem could resolve autonomously, so that climate policy instrumentations – new or old – would not be needed. Several ideas have been presented, each contributing little alone but possibly effective in combination. Existing environmental policy instruments together with key regulations, subsidies, and additional actions by cities to address local concerns might achieve the deep emission reductions required to meet the climate goal. Some of the mechanisms that have been proposed in support of this scenario are summarized below.

- The use of fossil fuels will fade out by itself due to ever cheaper renewable electricity and low-cost, electricity-using systems. Electric cars, for example, will become cheaper than combustion driven vehicles. The end of the fossil age is inevitable.
- Energy efficiency improvement is possible to a very large degree in virtually all energy applications, aided by new information technologies and big data. Self-driving cars will also reduce transport volumes and emissions. Energy efficiency could substantially reduce emissions.
- The circular, low primary materials economy will reduce emissions substantially. By closing material circles, energy requirements will also be reduced substantially.
- Co-benefits will create low-emission cities and regions. Air pollution will be controlled locally for local reasons, driven by local concerns. Fossil combustion in cities then has to stop. Such co-benefits will create prime reasons for reducing fossil emissions substantially.
- Firms and consumers will assume moral responsibility for their contributions to climate change and actively phase out most fossil energy applications. A cultural change will create significant shifts away from fossil energy use.

Emissions could diminish in the near future as a result of such autonomous developments, aided by already-existing instruments for climate and energy policy and broader environmental instruments such as the clean air instruments of old. Is there an instrumentation problem? Each line of optimistic reasoning will be investigated in turn, and the pessimistic as well: might climate policy have adverse climate effects?

Is the end of the fossil age in sight?

"The Stone Age did not end for lack of stone, and the Oil Age will end long before the world runs out of oil" is a quote from Sheikh Yamani, the former oil minister of Saudi Arabia. Is the "fossil age" coming to an end for reasons not linked to climate problems? Stone utensils were replaced by metals, which were cheaper and better than stone. Could fossil energy systems be replaced by other primary energy systems that are cheaper and better? Direct solar heat and photovoltaics, as well as derived solar energy, including hydro, wind, wave and geothermal energy, are each nearly endlessly available relative to economic requirements, and certainly in combination, along with nuclear fission and fusion. If the costs of non–fossil energy production decrease enough, fossil fuels may very well lose their edge in the energy market. Optimism is in the air: solar PV is assumed to outcompete fossil-based electricity, large firms are starting to reduce emissions voluntarily, and some cities are taking responsibility for climate policy for primarily local reasons. Could such developments snowball in the near future? Fossil fuel use has substantial environmental effects in addition to climate effects. As these other environmental effects are more local, they are more susceptible to local and national policies without requiring international coordination. A co-benefit of local and regional pollution control could be a substantial reduction in fossil fuel use. Banning polluting fossil fuel vehicles from cities could force fossil fuel–powered cars off the market. Clean cars are more expensive to produce, as the Volkswagen Dieselgate investigation demonstrated. But by raising the costs of fossil fuel energy, such other policies would combine with the first mechanism, that of cheaper alternatives to fossil energy systems. These alternatives, then, could outcompete fossils in the marketplace. Is this a realistic scenario?

The answer is a definitive "no." Intermittent wind and solar require costly backup over days and seasons and over low-wind periods, which in many regions may last up to several weeks. Emissions control of local pollution resulting from electricity production and transport can be reached by using focused technologies. This was achieved in Western countries in the 1970s and 1980s for sulfur and nitrogen emissions primarily at a national level, not just locally. The regulatory burden on cities would be very heavy, however, and the means of regulating emissions, in terms of instruments now available, is hardly adequate to the task. National policies would then be more effective and efficient, although again hardly affecting fossil fuel use, as such policies did not restrain their use in the past in the richer OECD countries.

Some firms may cater to a climate-sensitive audience. But the market for environmentally oriented products is small, and buyers are mostly focused on issues that are not primarily related to emissions reduction, such as achieving a circular society focused on sustainability, and banning plastic bags. Healthful foods and support for nature are key issues involving agricultural production. However, they compete substantially with climate in an ethical/ideological sense, and also materially. Organic food, for example, competes with other land use more than intensive agricultural production. Green tourism locations tend to require long-distance flights and may also disturb local ecosystems.

Firms could invest in lower emission technologies and markets based on green investor preferences. Most investors' prime motivator would be their return on investment. Being content with "1% less," for example by going from a 7% to a 6% compound interest rate, would reduce the income from a lifetime pension fund payment substantially, in the order of one third. But such decreases cannot easily be accepted, as pension funds are already under pressure due to aging of societies. Additionally, firms willing to sacrifice some profits for the common good might be bought out by those whose priority is higher returns. Of course, proactive actions taken by firms can be very useful, if only for technology development and market creation. And such firms may be smart in expecting adequate climate policy to come soon. But that is not equivalent to autonomous emissions reduction. Such action is limited to niches, with the fossil fuel supply remaining fully intact at current or decreasing costs. Energy prices have historically tended downward, with extreme peaks in only two periods of exceptionally rapid economic growth: the OECD spurt from the 1960s on, and the China spurt beginning in the 1990s (Zellou & Cuddington, 2012). With even limited reduction of demand, prices may drop substantially, slowing down any emission reduction.

Is energy efficiency policy the key to emission reduction?

Improved energy efficiency is widely seen as a most important independent factor entering into climate policy. Emissions reduction of CO_2 can thus be seen as a co-benefit of such energy-efficient policy. The IEA affords a central role to energy efficiency in all of its close-to-2-degree scenarios (IEA, 2016). It has formulated 25 policy recommendations on energy efficiency that could reduce climate emissions by nearly 20% in 20 years (IEA, 2011), a substantial long-term contribution. Similarly, improved energy efficiency plays a key role in most IPCC scenarios (IPCC-AR5-WGIII, 2014), including Representative Concentration Pathway 2.6 (RCP 2.6) scenarios that show a reasonable chance of staying below a 2-degree temperature rise. The EU has issued an energy efficiency directive with targets for member states and also certain obligatory instruments, such as a minimum number of dwelling renovations per year for each member state.

An increase of energy efficiency at a macro level requires decoupling energy use from economic growth, or the total primary energy supply from GDP. A relative decoupling that provides a reduction of energy use per unit of GDP is substantial.[7] But that reduction is fully negated by rising GDP, leading to additional energy use. Absolute decoupling has not occurred historically: energy use is rising relentlessly, as it has been for centuries, and increasingly so since WWII. Instruments for increasing energy efficiency are scarce and tend to be micro level and technology specific. Micro-level policy instruments, if successful, do not necessarily increase macro-level energy efficiency. Energy efficiency of residential space heating, for example, is measured in terms of energy use per square meter of floor (IEA, 2017a). The measured efficiency increase in most developed countries is counteracted, however, by an increase in floor space, which is closely connected to GDP development (Moura, Smith, & Belzer, 2015). The forced refurbishing of dwellings and strict standards for new construction are now major instruments in the EU energy efficiency directive (EU, 2012). However, these investments require energy use. This will produce a net increase of emissions initially, and thus contribute to a *decrease* in energy efficiency. Only later will there be pay-back, as energy use in dwellings drops – but this could very well take decades. Finally, consumers react quickly to lower energy costs with behavioral changes, such as by increasing heating and ventilation during colder periods and increasing cooling in hotter periods. When energy regulations are truly successful at a micro level, the costs of heating will go down, leaving consumers with income to spend on other expenditures, which also require energy, as in increased air travel and high quality energy–intensive food.

Some of these effects are covered in the analysis of rebound mechanisms. How several rebound mechanisms can decrease the net effect of technical improvements depends on the specifics of the situation analyzed, and the method of analysis used, as shown in a survey of 42 rebound studies (Font Vivanco & van der Voet, 2014). Detailed studies on transport innovations and on electric vehicles, in particular, show similar high and diverging rebound outcomes (Font Vivanco, Freire-González, Kemp, & Van der Voet, 2014; Font Vivanco, Kemp, & van der Voet, 2015). Failing to take account of rebound effects induced by necessity will lead to incorrect conclusions. Similarly, it is often assumed that added renewables, a micro-level development, actually *replace* fossil fuels by an equal amount at a macro level, with emission reduction linked to that replacement. This assumption does not generally hold, however: renewables add to the total supply of energy. They can replace only a part of the fossil fuel supply, as they make fossil energy cheaper through normal market mechanisms, increasing its use elsewhere. One example of not recognizing this micro–macro level discrepancy can be seen in a publication by the German Environment Ministry (Umweltbundesamt). It claims that reductions realized at the micro level of fossil energy through replacement with renewables production will be followed by the same reduction at the macro level (UBA, 2015, pp. 36–37).[8] In the same vein of reasoning, the closure of nuclear installations in Germany should be counted as *added* fossil use and emissions, which is the case, necessarily, to some extent as well. Furthermore, decentralized wind and solar electricity replace fossil-based electricity from the grid, which falls under the EU-ETS. Hence, the ETS tradable emission permits brought on that market will induce a lower permit price, allowing others to emit more. Such indirect effects are diverse and quite unavoidable. In terms of climate policy, it is ultimately the macro systems level that counts.

With successful micro-level energy and climate policy, the micro–macro divide still remains. Rebound mechanisms taken into account may be stretched to ultimately include cultural mechanisms and policy reactions for the long chains of mechanisms that may be involved (Huppes & Ishikawa, 2009). Some of the main reaction mechanisms analyzed include demand response, income effect and induced general technology development (Sorrell, 2009). But modeling of uncertain and conditional broader developments will remain elusive. Finally, it is notoriously difficult to make energy efficiency increases the focus of successful policy (Saunders, 2015). Even if they could lead to a degree of absolute decoupling, such improvements cannot reduce emissions by the required minimum of 90%.

Summary of energy efficiency in climate policy

- A confusing multitude of *energy efficiency* concepts exists at both the macro and micro levels.
- Energy efficiency improvements at a micro level do not directly correspond to improvements at a macro level due to conceptual difficulties and, more fundamentally, indirect mechanisms, as rebounds.
- Taking into account the life-cycle effects of choices at a micro level seems most sensible for creating real-life micro level improvements. Well-designed measures will reduce energy per functional unit, such as result from the *Life Cycle Assessment* (LCA). For long-lived products such as buildings, there may well be additional energy use and emissions over the short term of decades.
- Increase in efficiency is the motor of economic growth, with energy efficiency as a key contributor.
- Macro-level energy efficiency is a measure of society's performance. It can be defined as economic output per unit of energy use; that is, per unit of GDP per gigajoule (GJ). The inverse is energy intensity, the amount of energy per unit of GDP.
- Reducing energy intensity overall will also squeeze out all non-fossil sources, slowing down their introduction.
- The goal of substantially speeding up energy efficiency improvements at a macro level has proven elusive from a policy perspective, making it hazardous to base climate policy on this energy policy domain.
- A focused, pure climate policy approach seems highly preferable, leaving possible energy efficiency increases as a coincidental co-benefit for energy policy.
- These conclusions are in clear conflict with statements of the IPCC and the IEAA in policy proposals (IEA, 2016; IPCC-AR5-WGIII, 2014).

Energy efficiency policy might contribute in some uncertain way to climate policy, but never to the extent that climate policy becomes superfluous.

Can the climate problem be solved by moving toward a circular economy?

The shift to a circular economy focuses on resources used as materials in products. Many studies have looked at how the circular economy might develop, with recycling/

Table 1.1 Primary production and recycle/reuse rates for main resources

Resource	*Primary production (Mt)*	*Extracted value (B$)*	*End-of-life recycling*	*Recycled content in new*
Fossil fuels	12,000	5,000	0.0	0.0
Iron	1,260	390	0.7	0.4
Aluminum	60	100	0.6	0.4
Chromium	15	60	0.9	0.2
Copper	12	90	0.5	0.3
Zinc	10	25	0.4	0.3
Lead	5	10	0.7	0.5
Manganese	3	6	0.5	0.4
Nickel	2	24	0.6	0.4
Gold	0.0018	100	–	–
Tantalum	0.0008	0.1	0.2	0.1

Source (metals): http://www.resourcepanel.org/reports/recycling-rates-metals.

Note
B$, billions of dollars; Mt, megatonne.

reuse rates for such materials becoming as high as they now are for iron and copper. However, from a resource perspective the most noncircular resource is fossil energy, which is more than 99% noncircular. It is noncircular by necessity; extracting the energy from the source product is the aim, and CO_2 emission is a byproduct of that process. Also, fossil energy is the resource dominating all other resources coming into the economy, both by volume or mass and by value; see Table 1.1.

Fossil fuel energy required for primary materials production is generally much higher than that required for secondary production – although there are boundaries. Collection and processing of diffusely used final products may require more energy than virgin production. Nonetheless, there are many good reasons to aim for circularity as in metals, including energy and climate reasons. Most non-fossil energy sources are more materials intensive than fossil sources. Shortages in their primary supply may limit the rapid transition to a non–fossil fuel society. The main impediment is however a lack of clarity in climate policy, delaying the necessary investments (de Koning et al., 2018). Increased reuse and recycling may alleviate such supply problems, even for an already highly recycled metal such as copper (Elshkaki, Graedel, Ciacci, & Reck, 2016; ICSG, 2017; Schipper et al., 2018). For fossils, however, circularity means not using them for energy, or getting their carbon back into geological storage, if taken out at all. That is the main goal of climate policy.

Resolving the noncircular nature of fossil fuel extraction and use does not replace direct climate policy but must be the core of climate policy. The main options are to not use fossils and instead leave them underground, or to transport the CO_2 that is released when they are used underground to avoid emissions. Carbon capture and storage (CCS) hardly counts as circular, but it can be a part of climate policy, as it is in many scenarios for meeting the climate policy goal of limiting temperature rise to 2 or preferably 1.5 degrees.

Increasing the circularity in the use of material resources may reduce fossil energy use required in primary production (Graedel & van der Voet, 2010). Most metals are already substantially recycled by the end of their useful applications. For stocks that are being increased, as with iron, aluminum, copper, zinc and nickel in the built environment and other infrastructure, recycling will come later, in many cases well after the half-century mark, when under successful climate policy their production will have to have been decarbonized. The production of iron and steel dominate CO_2 emissions in global metals production for energy use but also chemically, in the reduction of iron oxide to the metal iron. Currently, iron has a very high rate of reuse and recycling, so increased circularity is not very relevant as a climate policy option in this regard. However, technological change in production is highly relevant, using options other than carbon for reducing iron oxide (for example, hydrogen). Overall, circularity can only play a limited role in reducing climate emissions, but where technically possible and economically attractive, such options may be used advantageously. Thus, climate policy contributes to circularity, not the other way around.

Can the climate problem be solved as a clean air co-benefit?

Most climate modeling, as described in the recent AR5 IPCC survey, depicts an ominous future if more stringent policies are not developed. Following a business-as-usual course will lead the world to a 6- to 8-degree Celsius temperature increase with high probability and an increasing likelihood of disastrous runaway effects. Is there a chance that these models are wrong, and missing the autonomous developments that can already be seen? Seemingly most autonomous is the electric car revolution, sparked by improved batteries. This "Tesla revolution" is a revolution in battery-driven cars and home batteries, which provide decentralized storage possibly in combination with decentralized renewables production (often solar PVs). That revolution is ongoing, and not limited to Tesla; China is the largest market for battery-driven cars. But what does "autonomous" mean when electric car sales depend on support by governments, using subsidies and tax exemptions and also fixed quotas in fleet standards, as in the United States, the EU, and China? Home batteries may be attractive for delivering an uninterrupted supply of electricity if a central supply source is unreliable and a well-developed electricity market with variable prices is lacking. However, where large-scale reliable grids develop along with open, real-time, variable-priced markets (discussed later in Chapter 5), secondary production of energy by means of home batteries will quite certainly *not* develop, as this is one of the most expensive storage options available. In such a well-developed market, rooftop PVs will be an expensive option in moderate climates such as the northern United States and northern Europe – unable to compete with southern PVs and concentrated solar heat electricity, produced at less than half the lifetime costs and with a more reliable supply year-round. For example, in the United States, PV electricity costs before subsidy are around \$0.06/kWh in the sunny southern states, \$0.10/kWh in the northern states, and much higher in Alaska, according to a Bloomberg report.[9] But even in tropical and subtropical areas the sun does not shine at night, and hardly at all on some cloudy days, requiring costly backup capacity. Without active climate policies, large-scale markets for electric cars, decentralized PV and battery storage will not be likely to develop at the scale and speed required to have an effect on climate change.

Will new climate technologies create jobs, driving climate policy?

It is often said, and widely believed, that renewables will create jobs. This is certainly true at the micro level; solar cells and wind turbines must be produced, installed and maintained – and each of these activities requires human labor. However, employment in the aggregate in not based on specific production technologies; instead, it reflects the overall volume and structure of demand in relation to the volume and structure of labor supply. Employment as a macro-level variable results mainly from two basic policy variables. Total demand results from macro-economic policy as it relates to business cycles and is independent with respect to climate policy. Next, structural unemployment is based on the mismatch between the supply of specific labor and the demand for it, and on frictions such as not starting a new job directly. Education, including lifelong learning, and industrial policies to match labor demand to supply, are major options for public policy, but not for climate policy. Specific micro-level employment measures would result from any increase in a specific expenditure, always at the cost of other expenditures. If new types of employment and schooling or training would be required, climate policy could even increase short-term friction unemployment as any innovation could. To deal with such developments, policy domains other than climate policy are already substantially in place; of course, these differ between countries and are open to improvement. Thus, employment creation as a co-benefit of climate policy is a dubious argument at best. Emissions reduction as a co-benefit of employment policy is even more dubious.

Could all non-climate developments combined lead to near-zero emission?

Specific micro-level energy efficiency policies have a limited effect on total energy use and emissions, if any at all. More circular economies would hardly contribute in any quantitatively relevant way on the path to near-zero emissions. Aiming for emissions reduction as a co-benefit of local pollution prevention seems irresponsible to the victims of air pollution now, as proven technologies and instruments for reducing local pollution are available today and have been for decades. Employment policy does not have a relationship to climate policy, except a negative one, in contributing to economic change. There may be some limited co-benefits, but even combined, these solutions to the climate problem would fail to meet the Paris Agreement goals. Conversely, some forms of climate policy could induce increased energy efficiency in terms of final energy use efficiency, and certainly in terms of primary energy efficiency increase, because primary electricity production has fewer losses on the road to final energy use than fossil energy. Additionally, electric power at end use, as in motor vehicles, is substantially more efficient than gasoline for combustion engines. Thus, there are some co-benefits from climate policy to energy policy, but almost none the other way around.

The Green Paradox: could expected climate policy increase emissions?

Effective climate policy reduces the long-term resource rents of fossils producers. Expecting effective climate policy, they would increase production now. That is the Green Paradox (Bauer, McGlade, Hilaire, & Ekins, 2018; Jensen, Mohlin, Pittel, & Sterner, 2015). However, if acting in a well-organized oligopoly, as is somewhat the case with the Organization of Petroleum Exporting Countries (OPEC) and its allies, it would be logical to constrain production, in this case of oil, to receive higher prices now. Oil and other fossils have a short-term inelastic demand. Decreasing production thus increases income while reducing costs. In this instance, the Green Paradox seems far from applicable. And, in fact, the studies of Jensen et al. and Bauer et al. find no empirical evidence for the Green Paradox to be a relevant mechanism due to climate policy. Countries with the highest fossil reserves would be hit hardest by climate policy as their resources would remain underground (McGlade & Ekins, 2015). That is an economic reason for them to oppose effective global climate policy, and is particularly relevant in poor but fossil-rich countries.

Discovering and creating the future

Approaches to governance and instrumentation are relevant to more than the setting of goals, targets and the means to reach them. They are also affected by and must respond to broader cultural, political, economic and technical developments in society, in individual countries as well as in the world at large. Feasible options are substantially restricted. The options and restrictions are partly known today and are also substantially uncertain, unknown and variable within broader societal developments. With the 2-degree/1.5-degree climate policy goal stated and the global emission reduction tasks clear, the first question is how to deal with the knowns and unknowns, and with the certainties and uncertainties between now and the low-emission future. Dealing with uncertainties is a basic issue in envisioning long-term policy, and is closely related to the level of detail in the analysis. Solar energy, in the aggregate, is a given. But at a more detailed level of technological specification there is direct solar PV, low temperature solar heat, and concentrated solar heat. For each of these main options very different technologies exist – each, again, providing different and substantially unknown options for further technology development. How can the knowns, uncertainties and unknowns be incorporated in instrumentation design? They are part of discovering and creating the future, finding out what is possible and attractive. Models of a different kind, including semi-quantitative and qualitative ones, will all play a role that is only minimally amenable to technology-specific instrumentation design. The two political philosophies differ substantially in how detailed projections must be.

The future: prediction, projection and backcasting, and discovery and creation

Several knowledge domains guide us in knowing the future – as modeled prediction or active creation, with partially predictive knowledge in between – in an ongoing process also known as *discovery of the future* (Kramer, 2017; Wells, 1902/1913). Predictions, or more modestly, projections, are possible futures based on knowledge of the current state and relevant mechanisms and relationships, on assumptions regarding more open-ended developments in policy, technology and culture, and on at least some knowledge of what is missing. Creating the future is possible only within the boundaries of the possible. Prediction, discovery and creation are not completely different from one another, but their respective starting points differ. All three are, by necessity, based on assumptions, because full knowledge of the future is not possible. Especially for long-term investigations, assumptions become dominant in determining outcomes. Unreasonable assumptions make outcomes irrelevant. But when is an assumption or combination of assumptions reasonable, and when is it not?

The IEA 2060 2-degree scenario (2DS) serves as an example of the difficulty in answering this question, as discussed later in this chapter (see also Table 1.3). The 2DS assumes a very substantial increase in energy efficiency at a macro level. The total primary energy supply barely rises, even though the OECD expects the global economy to grow by over a factor of three. Historically, growth of GDP and energy use have tended to go hand in hand, with only some relative decoupling in recent decades. Efficiency increase, not energy efficiency itself, is the mother of economic growth. To achieve absolute decoupling and realize economic growth with decreased total energy use, a substantially new energy efficiency policy will need to be developed. The focus then shifts to how such an energy efficiency policy linked to growth and to decreasing energy use could be developed. If the possibility of making this historical break in empirical relations is not seen as highly feasible, a substantially different scenario emerges, with very much higher renewable energy use – the fail-safe scenario (FSS) – summarized in Table 1.3.

IEA assumptions about biomass energy can also be questioned as there is a significant body of literature indicating that biomass energy may indeed produce energy but in larger volumes will produce higher climate-changing emissions per unit of electricity than natural gas or even coal-based electricity (Stephenson & MacKay, 2014). Similarly, the expanding role of nuclear energy assumed in the 2DS would be jeopardized by a few Fukushima-type accidents. Were several such events to occur, would anyone want to have a reactor in their backyard. Differing assumptions lead to very different outcomes, and to different requirements of climate policy and its instrumentation. Modeling the future is essential for wise climate policy instrumentation design, but by necessity assumptions remain uncertain. Avoiding the risks that a projected future might not come to pass means choosing a *fail-safe* approach and having policy instruments ready to deal with real possibilities such as promising low-emissions technologies that do *not* deliver. Therefore, the FSS shows offers only a modest reduction in energy intensity and has an *eightfold* higher requirement for solar and wind energy.

Models for climate policy instrumentation support

In a very real sense we make our future. Individuals and private and public organizations at several levels of government make myriad small-, medium- and large-scale decisions every day. Some feedback is derived from past experience: what went right, what did not, and what to do about it. Bright new ideas may present themselves. Expectations may change and ideas and goals may change. How can climate policy and its instrumentation help to create a consistent and reliable future? There are very different ways to look into the future. In the climate and energy domain, operational quantified models can be grouped into three main types: prospective integrated assessment models (IAMs), future scenarios based on input–output (IO) models, and partial optimization models, especially regarding energy and transport.

Integrated assessment models

One can predict or preview consequences of specific actions, or broader sets of adjoining actions, all then embedded in general trends that take account of main mechanisms in the full economy. IAMs first specify the results of climate-changing emissions. Next, they may go beyond emissions into climate effects and other environmental effects, with feedback loops to the economy. Such prospective models are broadly used to support climate policy and, closely related, energy policy. They also address regional pollution problems such as those caused by sulfur (SOx) and nitrogen (NOx) emissions and land use change. They combine knowledge from the past with statistical relationships as then existed, adding certain behavioral mechanisms to better reflect consequences of the choices to be analyzed with the model. Choices such as emission pricing or banning of coal are entered as exogenous changes in the model. To address climate feedback to the economy, the time frame must be extended, covering developments to 2100, and even beyond. Some longer-term restrictions can also be modeled, including land use, land use change, and depletion of nonrenewable resources. Since the models build on the past, and in the behavioral domain include a few primarily economic market mechanisms, they can cover new and substantially different technology and market developments only as scenarios. These models have been developed systematically by improved historical analysis and by grouping them in categories that have been referred to as shared socioeconomic pathways, or SSPs (Nakicenovic, Lempert, & Janetos, 2014). SSPs now are grouped into five main scenario pathways, each aligned with a global–regional development direction (O'Neill et al., 2015; Riahi et al., 2017).

The long-term validity of prospective IAMs is therefore limited: they show the world as it could develop, learning from the past, as sophisticated conditional extrapolation. Further improvements of input data are possible, reducing data uncertainty, but the validity of the models is not easily improved. Most IAMs still operate at a highly aggregate level, and assume near-ideal markets in the behavioral parts of their scenarios (IPCC-AR5-WGIII, 2014). Again, how reasonable are such assumptions with regard to data and mechanisms? Very-low-emission scenarios assume large-scale negative emissions in the second half of the century, based on large-scale low-emission biomass production with effective CCS, for example. But Stephenson and MacKay (2014) and Anderson and Peters (2016) contend that this assumption is not well-founded, if not a near impossibility. Reasonable technical assumptions may well indicate the contrary. Negative emissions may also result from DACS – direct air capture and then storage (Lackner et al., 2016). Additionally, the very large-scale funding required could well prove to be an unsurmountable sociopolitical obstacle, as the costs of this form of CCS will generate no proceeds to those incurring them. An IAM example linked to more specific technical and economic measures in climate policy is provided by Deetman, Hof, and Van Vuuren (2015), based on more specific technology assumptions, closer to a known current reality.

IO futures scenarios, with backcasting

Another modeling type starts with a detailed future state to be reached (or avoided). A zero-emission society will have an economy much adapted for emissions reduction. It can be sketched using detailed input–output relations and technical and behavioral assumptions. Large IO databases have been developed, and modeling includes waste management (Nakamura, Nakajima, Kondo, & Nagasaka, 2007), global resource use (Duchin, Levine, & Strømman, 2016) and economic technology and scenario analysis (Suh, 2009). The role of specific technologies can be added using LCA of the technologies involved, as a hybrid IO-LCA model for depicting a low-emission future society (De Koning, Huppes, Deetman, & Tukker, 2016). In specifying a zero-emission society, one may further assume a dominant role for primary solar energy and a fully electrified transport system, for example. Through technical specifications developed on the basis of prospective technology analysis, the supply chains can be specified and built into the IO sector framework, showing overall results. The sectoral and technology detail in IO models can be much more detailed than IAMs can handle.

In addition to providing more detailed specifications of the future functioning of society, IO modelers can use backcasting to show what is required to get there. Using a future date, such as 2060, researchers would start from the premise that anything that then exists must have been produced before that date. Lithium battery volumes can be set, with the buildup of lithium on the route to their very broad application specified through backcasting. Similarly, total copper use can be specified, indicating how much production must increase, accounting for stock increases and recycling. Connections can be made with IAMs. Sectoral detail of IO models, up to hundreds of sectors for each of up to a hundred countries, can be aggregated to the level of detail covered by IAMs, linking to their broader, more aggregated economic mechanisms, which are limited or absent in IO models. These IAMs, in turn, may increase their sectoral detail, due to rapid developments in computer processing power. Backcasting models and prospective models may thus meet when taking into account mechanisms in society as modeled in IAMs, if they happen to depict the same future. Prospective IAMs and backcasting models for climate policy both cover the full global economy, and hence all emissions that contribute to climate change. The danger of both approaches is that they tend toward technological and economic determinism, and to ever more detailed policy actions to indeed create the future as one of the detailed specifications agreed upon: the best available.

Partial optimization models

Quantified models covering only a part of the economy are also available. They can give insight into how specific measures may impact specific domains. If, for example, coal and natural gas-fired power is regulated by increasingly stringent emission standards per kilowatt hour, how will other primary production develop? What about secondary production? And how would this relate to the market structure, with links to demand reactions? All further model-external specifications are based on experience and on outcomes of other models, or on assumptions that often dare to achieve near-zero emissions as an explicit boundary condition. Energy models, transport models and many more abound; the recent IEA model (IEA, 2017b) is one important example broadly incorporating much of the recent energy modeling efforts. More partial models can give relevant input. For instance, if self-driving vehicles were to become standard, private cars would become less interesting. What would this mean for the volume of the future fleet of personal automobiles? Next, a further partial model could indicate how public transport would be impacted or could actively adapt. Interrelationships in the broader economy would not be covered in any detail. But within these constraints the model itself can be very transparent and relatively simple to understand. Feeding such information into IO models covering the full economy can still be done quite transparently. Several such partial mechanisms may then be introduced into related IAMs, thereby improving their quality and detail. However, the interrelationships then included would make it more difficult to attribute effects to single mechanisms. Complex models with feedback loops such as IAMs tend to be opaque.

How climate policy might or should work out in models

The three main types of models supporting the design of climate policy each have advantages and disadvantages. Partial models are clearer but miss out on overall effects. IO models can be very detailed with regard to technological specifications but lack behavioral mechanisms. IAMs cover behavioral mechanisms in a general way but are inherently complex and opaque. Together they constitute what we can know quantifiably about the climate future, with indications of how society can be steered in a desired direction. But there is more knowledge than that contained in quantified models.

Back-of-the-envelope, semi-quantified models

The simplest models are so-called back-of-the-envelope models. They cover partial relationships between technologies and society and are closely linked to simple assumptions. If a CCS that is 80% effective is applied to all coal-fired power stations, how much would this reduce global emissions, without reckoning with indirect mechanisms? At what price levels for electricity and gasoline would electric cars become the cheaper option, assuming similar production costs and performance? If East–West connections are established in the electricity grid across several time zones, how much of the daily and weekly variation of intermittent electricity supply and demand would be cancelled out? What differences in electricity price level are required to make the purchase of batteries for demand stabilization profitable? If well chosen, such "little" models can provide insight into specific instruments or incentive requirements.

Modeling of single technologies

Most key technologies for very-low-emission futures have a substantial degree of uncertainty about them, in a combined technical-economical and sometimes social sense: they just may not happen. What is the expected reduction rate of emissions from coal-fired power stations by introducing CCS, all costs considered? What are lifetime emissions per kilowatt hour of solar PV, geothermal electricity, hydropower and nuclear fission? The answers may diverge substantially with respect to circumstances and assumptions, as shown in Table 1.2. The figures there include upstream emissions. These upstream emissions will become lower by 2050 as many will have been reduced by then; on this point, refer to the conceptually innovative paper by Hertwich et al. (2015). In a zero-emission society, the climate emissions of upstream processes for renewables and nuclear power would be zero because all energy used would be zero. The variability of emissions per technology also implies that tools for analysis, such as marginal abatement cost (MAC)–based ranking of technology choice, is already a dubious approach in the short term (Kesicki & Ekins, 2012). This certainly holds for long-term choices. Optimization models using lowest cost options for low-emission energy cannot avoid these high uncertainties, and cannot easily be used for technology selection, but they can give rough indications (Solano & Drummond, 2014). External checks are required to determine how reasonable outcomes are. If economic optimization would lead, for example, to nuclear energy becoming dominant in energy supply, this may be seen as unreasonable as for technical or political reasons. The nuclear production volume then can be reduced by setting a higher expected cost price, or by some other quantitative boundary set "by hand" in the model. Simple technology models may cover such assumptions more directly, indicating their relevance in reasoning toward specific technologies.

Table 1.2 CO_2 equivalent emissions per MWh of primary electricity production

Technology	*Min*	*Median*	*Max*
Currently commercially available technologies			
Coal (PC)	740	820	910
Gas (combined cycle)	410	490	650
Solar PV (utility scale)	18	48	180
Solar PV (rooftop)	26	41	60
Concentrated solar power	8.8	27	63
Geothermal	6.0	38	79
Hydropower	1.0	24	2,200
Wind (offshore)	8.0	12	35
Wind (onshore)	7.0	11	56
Nuclear	3.7	12	110
Pre-commercial technologies			
CCS – Coal – PC	190	220	250
CCS – Coal – IGCC	170	200	230
CCS – Gas (combined cycle)	94	170	340
CCS – Coal (oxyfuel and CCS)	100	160	200
Ocean (tidal and wave)	5.6	17	28

Source: IPCC-WG3 (2014), Annex 3, p. 10.

Note
CCS, carbon capture and storage; IGCC, integrated gasification combined cycle; PC, pulverized coal; PV, photovoltaic.

Using information from models for instrumentation design

Planning & Control instrumentation must deal with specific technologies, to specify them and induce them, and to remove what does not fit. Technology-specific modeling, therefore, is a requirement for effectiveness of both instrument-technology combinations and overall instrumentation. Effect-oriented instrumentation for encompassing instruments may deal with groups of technologies, such as fleet standards in terms of CO_2 per car kilometer covering all personal automobiles and CO_2 emissions per kilowatt hour of electricity covering all fossil fuel power stations. Modeling such measures in IAMs or IO models may indicate the effectiveness at the full systems level. Boundary issues will often arise. Examples include combined heat and power and chemicals production in coal gasification and small trucks and sport utility vehicles (SUVs) adjacent to fleet standards for personal automobiles. Substantial market shifts may be induced by partial instruments. Smaller models covering a limited technology domain may bring such developments into policy consideration, then adapt the original instrument or add instruments to contain such undesirable shifts. As the unforeseen creates uncertainties, overshooting the emission reduction mark seems required, as one variant of modeling.

Basic uncertainties relate to nuclear energy, CCS, and biofuel – and to biofuel and CCS combined to achieve negative emissions. Inducing not-yet-existing technologies is one issue that is hardly resolved by modeling, and hardly resolved by technology-specific instrumentation. Accepting the essential role of renewables may be useful in setting targets. It does not, however, indicate which specific renewables to implement by some policy instruments. Modeling the role of biofuels, particularly their usefulness, is already too demanding to provide clear outcomes. Land is abundantly available when spreading productivity levels such as already exist to the rest of the world, and reducing meat consumption substantially to healthier levels. But can such changes be induced? Instruments to do so hardly exist. Next, other reasoning comes in, comparing land use of biofuels with direct solar conversion. Land use requirements for PV are very limited, with energy proceeds per hectare two orders of magnitude higher than for biomass energy (Pimentel & Pimentel, 2007, table 19.3, p. 262). Also, the highest electricity yields are in hot desert areas, where there is no competition with food production and nature. But the question there is, again: will it come, there, and which constraints are to be overcome? If expanding grids internationally is required but difficult, it may not be wise to reckon with large-scale desert PV. Modeling the diverging sets of uncertainties and assumptions may easily send clear outcomes of models astray. Is current biomass policy wise, with use of bioethanol and palm oil for car fuel and wood pallets for electricity production? Are negative emissions of bioenergy with CCS a feasible option, technically and politically? Some of these uncertainties can be specified in backcasting scenario models of the IO type. The FSS option, discussed in more detail later in the chapter, offers one still very simplified starting point.

Institutionalist instrumentation focuses on generic incentives and option creation. Technological specifics of models are not important for instrumentation. Uncertainties surrounding the success of new technologies are countered by the new options that may be successfully induced by an emission tax. In due time the tax may be adapted if overall reduction is too slow. A too-fast reduction is hardly possible; but performing better than modeled would be a nice present. With close to zero emissions, the tax hardly hurts, ultimately.

Summary of modeled knowledge

Relevant technical knowns must be taken into consideration in designing effective climate policy instrumentation:

- Renewables, including solar and wind energy, are available in practically unlimited amounts, but are not always close to their useful application, and are always variable and mostly nondispatchable.
- Nuclear, CCS and bioenergy may each play a role – even an important one – but are basically uncertain and cannot be relied upon. This holds true for large-scale geothermal power and solar synthesis gas as well.

It then follows that to remain on the fail-safe side:

- Instrumentation should be able to create fossil-free primary energy, as heat and electricity produced using solar PV, solar heat, hydropower and wind.

- Transport must be nearly fully fossil free, including aviation and shipping, where substantially new technologies will need to be developed.
- Heating of buildings should be nearly fossil free, based on emissions-free electricity and on solar and geothermal heat.
- Industrial processes using heat and power, including chemical processes, must also become predominantly electricity based.

The relevant socioeconomic high probabilities are as follows:

- The market economy is here to stay, for the most part globalized, with globalized culture and information exchange.
- Global economic growth will continue, possibly at an extreme rate of up to 4% per year during the next half century, higher than the OECD's best guess of 3%. At 3%, global GDP will nearly quadruple from base year 2015 by around 2060. At 4% it will rise nearly sixfold.

These are not predictions but realistic worst-case modeling outcomes from a climate perspective that must be dealt with in the strategic design of any instrumentation.

In what ways could the world develop beyond the models?

Most modeling relies on what has been accomplished in the past. Major transitions, regardless of whether intentionally made or not, are based on new relations that build on the old relations, but only partially. The semi-conductor and IT revolutions were certainly not foreseen, even though they were generated by old institutions. Long-term technical, economic and social trends in industrialization may now be bringing us a fourth Industrial Revolution (IR4). IR1 brought coal power and the mechanization of labor; IR2 brought electricity and transport and first-round automation, and IR3, IT and second-round automation. IR4 is now ushering in real-time cyber information and artificial intelligence (AI).

IR4 may again reduce, possible substantially, the human input required per unit of output, thereby increasing labor productivity and again rendering whole classes of employment superfluous, as all previous technology revolutions did and continue to do. That is the essence of economic growth. The driverless car is one clear example, which will arrive in some form soon. The handling of complex activities can be taken over by machines based on AI. Somewhat standardized intellectual tasks can be taken over by intelligent systems, as well. The nature of tasks is shifting even in the realm of academic research and education. How scientific documents and data are searched for and retrieved has already changed in the course of the past decade, making many library tasks superfluous and speeding up research significantly. Information overflow for the poor overloaded brains receiving unlimited information so quickly is an issue in this regard. AI will move in that direction as well, speeding up research even more, and requiring fewer researchers for improved results. Again, directions in AI may differ between hierarchically ordered specialization and broader integration, and may be more difficult to design in advance. Managed development by governments or large firms may dominate, or more evolutionary bottom-up processes may play a larger role.

A direct conceptual link can be seen to directions in climate policy development: planned development versus open development based on centrally developed incentives, options and constraints. Such broad developments occur mostly independently from climate policy instrumentation, and climate policy may contribute to such wider developments as directions in governance, linked to political philosophies and related governance views. A global, legally binding agreement on annually decreasing emission volumes per country must be closely linked to centrally planned instrumentation at the national and regional levels. Internalization of external climate effects, through an internationally agreed level of emission taxes among countries, would point in a very different direction for national climate policy instrumentation.

Would capitalism end?

One characteristic of wind and solar renewable energy is their near-zero marginal costs. The sun shines and the wind blows for free. It is nearly inevitable that they will play a substantial role in effective climate policy. A zero-marginal-cost society as envisioned by Rifkin (2014) would lead to zero prices in times of abundant production of the zero-marginal-cost product. Having most products available for free would herald the end of capitalism. Would a zero-emission society, as part of that development, indeed require, or lead to, the end of capitalism? If so, optimism about effective climate policy would certainly be difficult to justify politically with the powers that be. However, with short-term near-zero marginal costs, long-term marginal costs must still be covered by proceeds of the firms investing. Unless long-term total costs are matched by long-term total proceeds, investment will not occur, or will suffer a loss. This is the norm in economics, with near-zero short-term marginal costs now in hotels and public transport, for example. What is new is that winner-takes-all technologies have become dominant in ever more knowledge-intensive global markets: everybody wants only the best (Brynjolfsson & McAfee, 2014, chap. 10). The assumed best then takes over the market, as occurred with Microsoft's Office and Google's search engine. This is similar to the familiar problem of trends toward monopolization or, more generally, to oligopolies. Monopolies tend toward stagnation. With oligopolies, stable situations are less probable and outcomes less predictable, while behaving as a monopoly in any market domain is attractive for all firms involved but is mostly forbidden. The combined trends of zero-marginal-cost and winner-takes-all may well lead to one or a few firms taking over the new renewable energy market and reaping profits while underinvesting. A very different solution could be public funding and regulation of the monopolistic part of the new energy chain: the electricity grid. Such a broader public role not only links to the public good provision, as analyzed in welfare theory (Samuelson, 1955), but also does so for more pragmatic reasons – that is, desired results (Mazzucato, 2013), or for more normative reasons (Raworth, 2017).

Creative destruction will abound

The low-emissions revolution as part of IR4 would cause the disappearance of entire sectors of current industries, and many of today's occupations would similarly fully or partially disappear – Schumpeter's *creative destruction* (Schumpeter, 1942/1947). This happens in all rapid transformations, in any technology revolution. Employment in agriculture, which had a share of around 50% in Europe in 1900, has been reduced to well below 5% today. In China it dropped from 80% in 1965 to 30% in 2015, with substantial further decreases to come. However, climate policy will drive very little of the future disruption in the labor market. New groups of people will lose out, based on the type of job, age, and where they live in cities and regions. Action will be required by and for those involved in order to adapt to new evolving systems. New types of employment will be created, with education adapted to support it. This is a pressing task in an adverse social and economic climate, likely to be the case in formerly flourishing industrial areas, where such proactive efforts should be taken to prevent economic and social breakdowns. Firms will have to start up in as yet unattractive and possibly unsupportive surroundings. Successful climate policy may contribute to the destruction of existing industries and markets but with a limited share. Even a full shift from combustion cars to battery-powered electric cars would not be very consequential when compared with the ongoing reorganization of production in all car manufacturing and trends toward self-driving vehicles, and with similar development in most other industries.

To preserve governmental legitimacy, solutions to the economic and social disruption will be needed – not preventing the disruption, in the case of low-emission society to come, but accommodating it, which is possible in principle. At the beginning of the 19th century in England, the Luddites attacked weaving mills that were effectively destroying their home-based weaving jobs. They lost their battle. The loss of employment in agriculture has been much greater, and has continued over the long term. At first a burgeoning proletariat developed, controlled politically by the new middle and the old upper classes but more than 150 years later ultimately incorporated into the welfare society of the 20th century. Luddite-type bottom-up controls were ineffective from the beginning and are now nearly impossible. Winners will not necessarily be found at the same locations as the losers. Able to create their advantages elsewhere, they may obliterate the old technologies and destroy whole regions in the process unless the regions are resilient and can adapt to the new. The political consequences of disruption may be substantial and detrimental to climate policy, as well as social policy. They are not caused by climate policy in any quantitatively relevant way, however, nor can they be resolved by climate policy.

The information-industrial complex and climate policy instrumentation

Climate policy and its instrumentation need to be considered in this broader long-term context. The current governance approach is to seek effective instrumentation now: a course that involves setting targets and working toward achieving them, aided by all of the knowledge that is now available and will be developed in the future by the applied sciences. A dazzling amount of detail is required, which will also become increasingly manageable with new information and AI technologies. The alternate approach is more reflective about overall development within a framework of societal governance, with substantial risks related to new hierarchical power configurations evolving in business and government. These risks are only insignificantly related to climate policy. The choice of climate policy instrumentation relates to the problem of an ever-encroaching technocracy led by monopolistic global firms and by an issue-related, technology-focused political process. Detailed regulations will need to be rolled back to break the power of the *information-industrial complex* – a concept akin to but expanding beyond the *military-industrial complex* that U.S. President Dwight D. Eisenhower warned Americans about in the 1950s. In the climate domain, looser generic incentives and greater option creation are associated with that view of a more open societal governance. A more restricted reasoning toward this institutional approach is also possible, viewing it as most effective and efficient for reaching climate policy goals. Such a version of Planning & Control then would move toward Institutionalism, and away from detailed regulations.

Climate policy tasks for a near-zero emission society

The emissions budget

Reducing long-term climate change means reducing fossil CO_2 emissions over the long term. Reducing nature's biomass through deforestation has also contributed long term to rising CO_2 concentrations, although this reduction has declined over the past quarter century. Large-scale use of biomass for energy production could reverse that development, as old forest is transformed into productive forests and increased agricultural surfaces shrink natural habitats into ever-smaller areas. Of course, other greenhouse gases also contribute to global warming, especially methane and nitrous oxide, as previously noted. Current methane emissions contribute in the short to medium term to climate forcing, but contribute less to a rise in temperature in the longer term future. This allows relatively more time for policy development regarding methane, but not for the other greenhouse gases that are long lived, Concentrations of nitrous oxide, in particular, are rising and it will remain a potent greenhouse gas for many centuries.

A survey paper on the still-attainable emission budget for not overshooting the 2-degree Celsius temperature-rise goal clarifies various budget concepts and takes into account non–CO_2-elated emissions (see Rogelj et al., 2016). The global total Threshold Avoidance Budget (TAB) from 2015 onward is between 590 and 1,240 $GtCO_2$, depending on modeling types and assumptions (p. 251). The TAB calculates the amount of cumulative CO_2 emissions in a multi-gas emissions scenario needed to limit global mean temperature increase to below 2 degrees with a 66% probability. For a higher probability – never 100% – of remaining below an increase of 2 degrees the budget is lower. For the more ambitious task of remaining below a 1.5-degree temperature rise, the budget would be even substantially lower. As a mid-value for 2 degrees, and for ease of computation, a budget set at 800 $GtCO_2$ seems most appropriate working target. This amount is equivalent to about 20 years of CO_2 emissions at 2015 levels of around 40 $GtCO_2$ per year. Non-energy CO_2 emissions from land use and land use change are around 4 Gt per year and those from cement production are around 1.6 $GtCO_2$ (Olivier, Janssens-Maenhout, Muntean, & Peters, 2014). Fossil-related emissions were around 35 $GtCO_2$. Assuming over-optimistically that these other non-fossil CO_2 emissions can be reduced in similar percentages as fossil CO_2 emissions, a reduction path for the total can be sketched. At 40 Gt per year the 2-degree budget would be spent in 20 years. They then would have to fall to zero after 2035 – a complete impossibility. Spreading the budget out over 40 years using a linear reduction implies a cumulative reduction of 1 $GtCO_2$ per year, constituting 2.5% of 2015 emissions. This implies zero emissions by 2055 with a near-90% global reduction by 2050. This linear reduction of 1 $GtCO_2$ per year corresponds roughly to the more familiar compound emissions reduction rate of close to minus 6% per year. It would shift the rising trend of the past decade and half of plus 3% toward minus 6%, a significant 9% shift. The minus 6% corresponds to a reduction of 2.4 $GtCO_2$, starting directly in the first year – even more impossible than beginning the 1 Gt emission reduction now. Of course, such emissions reductions did not begin in 2015 and will not start immediately in 2020, barring economic disasters. Consequently, the emission budget has been overused, and later reductions will have to be even higher.

The policy instrumentation task here translates primarily into reduced fossil emissions. Other sources cannot, for the most part, be monitored in a detailed, quantitative way and hence need different, more complex instruments: fossil emission reduction is administratively relatively easy. Fossil emissions reduction translates into reduced fossil energy use, possibly mitigated by CCS, and for economic reasons will require a steep expansion of non–fossil energy use.

Instrumentation for resilience: the fail-safe scenario (FSS)

Expecting uncertain technologies to help reduce emissions in full creates a probability of not reaching the 2-degree, preferably 1.5-degree, emissions reduction under reasonable economic conditions. Other, more certain, technologies would have to be expanded if the uncertain technologies underperform. A fail-safe approach would not leave out uncertain, but possibly attractive, options from policy attention, but would take into consideration the worst case. The worst case involves a high-energy growth scenario – one that does not expect substantial reduction in energy intensity because of the lack, as yet, of energy efficiency policy. Opting for the fail-safe approach leaves out uncertain options such as CCS, large-scale biomass energy and large-scale nuclear energy. Only solar and wind seem to have no fundamental constraints. Which certainties and uncertainties, therefore, must be kept in mind for developing successful climate policy instrumentation?

The only certainty in the instrumentation design process is that fossil CO_2 emissions must be reduced to near zero. This is the normative starting point. This is in addition to the reduction of non-fossil CO_2 emissions and non–CO_2-related global warming emissions, which are not investigated here in any depth but are required as well. How the fossil CO_2 goal could be reached is substantially undetermined and undeterminable as long as technology development is undetermined, as is the rate of economic growth creating increasing demand for energy. Even the amount of fossil fuels compatible with this goal is not fixed, as greater CCS can allow for higher levels of fossil production to continue in the short to medium term. The composition of the mix of different fossil resources is also unfixed. Some coal gasification technologies, for example, may provide good options for cheap carbon capture. CCS is technically an option that is already available, but it is costly and will be increasingly so if higher capture rates are to be realized from industrial installations. In addition, its costs are merely added costs, which governments can avoid by leaving that task to others. Paying for CCS involves paying for the global collective good, without a return as is the case when paying for solar PV and wind turbines: a dollar paid is a cost for the payer with a return of a future lower climate risk that is negligibly small. This not only holds for firms but for most countries as well, with the exception of large countries such as the United States and China, the more loosely organized EU, and India, in future, which together produce the largest share of climate-changing emissions. Binding agreements would be required.

Without a doubt, energy is needed for the functioning of society. But how much is needed depends on many factors, partly on climate and other policies and partly on broader developments in technology and society. Overall, economic growth and the reduction in energy intensity determine the amount of energy used: GDP multiplied by Energy per unit of GDP is by definition the amount of energy used. Macro-level development is determined by meso- and micro-level developments. Self-driving cars may reduce traffic volume substantially, but they can also increase it. Long-distance, high-speed transport may become highly efficient if developed with a hyperloop system, or it may not. Such efficiency improvement may reduce energy needs overall, or even increase them through rebound mechanisms, as most eloquently worded in the Jevons Paradox, discussed earlier in this chapter. Remaining income effects may be spent in a high-energy manner, such as on long-distance aviation. During the 20th century and the two centuries prior, the net result of increased energy efficiency combined with economic growth was increased energy use, with only *soft decoupling* of around 1% to 2% per year in the past three decades. Some climate policy instruments may play a role in overall energy use, however, such as taxes on CO_2 emissions and other cost-raising instruments. The long-term, macro-level price elasticity of energy demand is on the order of one, meaning that if prices go up by 10%, the volume of energy use will be reduced by 10%. Japanese energy prices are nearly twice as high as those in the United States, and its energy intensity is less than half. Colder and (former) communist countries have a higher energy intensity. Interestingly, this empirical relationship implies that total expenditure on energy remains roughly the same

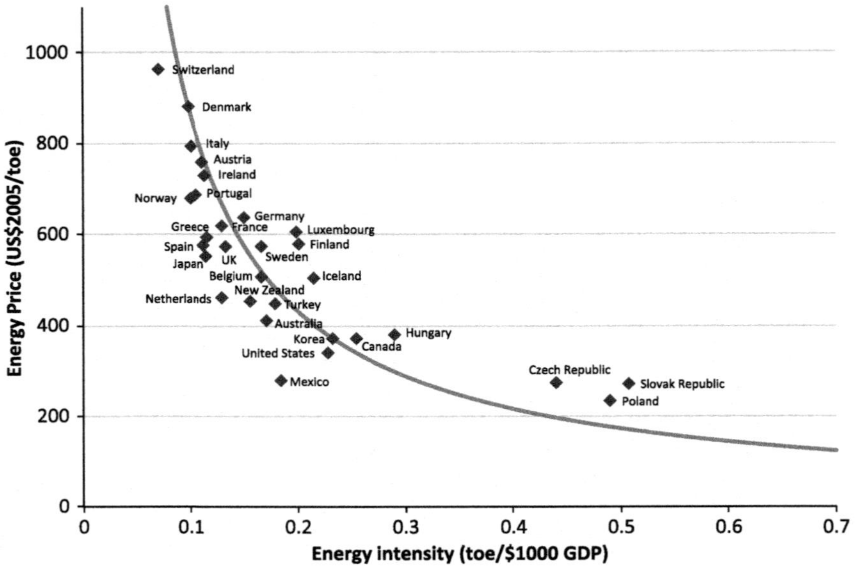

Figure 1.4 Average energy price and energy intensity by country, 1971–2012.

Source: Grubb et al. (2018)

at any after tax price level. Countries with higher taxes spend less on energy in real terms but have the proceeds of the energy taxes, which allow them to reduce other taxes, or to spend the proceeds on useful outlays that would not otherwise be funded. This is a relative certainty. The empirical analysis is depicted in Figure 1.4.

Uncertainties abound on the side of primary energy production. What role could bioenergy play? Technically it has a large energy potential. However, energy production per hectare of biomass produced for energy is only a fraction of that of solar cells (Pimentel & Pimentel, 2007, p. 262), with hence much higher land requirements. Even if attractive in terms of cost, a great deal of uncertainty exists with respect to the emissions of biofuel-based electricity production. Over the life cycle, such emissions could be high if used in large quantities, possibly higher than those produced by using natural gas or even coal (Stephenson & MacKay, 2014). Also, the potential volume of biomass energy production is limited due to land availability and competing requirements for food and nature. Current lifestyle trends also lead to higher meat consumption with rising income, requiring significant land use for agriculture. Some observers already see substantial net additional emissions, while others see a huge potential for bioenergy, especially if more people choose vegetarian diets, thus increasing the availability of land for bioenergy, or if ocean-based algae production could be made competitive and substantial. The possible amount of biomass energy ranges from zero to a substantial increase, with emissions ranges from much more to much less.

Next is the question of how much nuclear expansion can be expected by 2060? Planning for expansion and some construction has begun. But the occurrence of a few more Fukushima-type accidents would reduce that number to zero. The probability of such events occurring is not known, but it is certainly greater than zero. Also with current reactors under construction, costs tend to soar above planned levels, with very high subsidies required for commercial viability. Hydropower may expand, but within overall physical constraints in terms of reservoirs and rainfall. Also, how useful is hydropower in terms of climate policy? Tropical hydropower, in particular, may release such large amounts of methane that it constitutes a high global warming energy source (see Table 1.2). An option for recovering this methane for energy use might be found, although as yet this is highly uncertain (Ramos et al., 2009). Finally, there are also serious soft limits relating to protection of nature that may weigh more heavily in the future; this is already the case in Norway, where hydropower production is capped.

Many alternative energy sources are being explored that might or might not develop, such as the generation of electricity by mixing freshwater and saltwater, or beaming in solar energy from space or the moon. Such options should not be ruled out in advance, but they are also not guaranteed to deliver, and certainly not in substantial amounts anytime soon. The oceans could deliver tidal power, wave power and flow energy. The potential is huge but so are the costs. If costs decline below those for other intermittent primary sources, these alternative sources could end up playing a substantial role. But can such extensive development be expected? This hardly seems probable, let alone certain. The oceanic contribution may therefore remain limited. Geothermal energy has a huge potential. It will certainly develop in the most attractive locations, such as is now occurring in Iceland, for example. But in less attractive locations it is expensive relative to other non–fossil fuel options, and not something to be depended upon for large-scale energy production.

There is little doubt that wind power can be expanded substantially and that solar power can be expanded to near-unlimited levels relative to substantially rising human energy use. It is not certain which technologies will develop and in which mixes. Some forms of concentrated solar power (CSP) can store high-temperature heat for hours or even days, but could remain more expensive than other forms of solar power. Solar and wind are inherently intermittent, so that matching supply and demand becomes a substantial task with a significant number of renewables, possibly resolved through technical and institutional means, as by developing large, regional-scale, real-time, variable-priced electricity markets.

The result of all of these unpredictable uncertainties and a few certainties is that it is largely unknown exactly how the future energy system under the full restriction of 95% emission reduction by 2060 will or must look. This clearly is the case at the detailed, technical level, where, for example, at least four basic types of solar PV cells are on the market and more are being developed. What we know with certainty is that in 2060 the combination of solar and wind energy can technically supply any feasible amount of energy for society, as well as for rising demand beyond. However, this does not mean that climate policy should focus on solar and wind alone. Since all other options may be subject to severe restrictions, climate policy should open up the option of extreme development of wind and solar energy production, while not closing off the development of other potentially relevant and attractive options. The fail-safe option should be incorporated in instrumentation design in order to make it resilient. The amount of solar and wind then required is incredibly high compared with almost all scenarios surveyed by the IPCC. Because negative emissions, such as those based on CCS of biomass energy or extraction from the air, may not be realized, a near-full shift to solar and wind by 2060 should be a possible outcome of any resilient instrumentation. In the next section the FSS approach is quantified relative to the IEA 2DS approach. The amount of primary production of wind and solar electricity then rises from 17% in 2DS to 75% in FSS. There are many options for how to store and transport this intermittent nondispatchable energy for final consumption; similarly, there are many options for adapting demand to a given supply structure. Secondary hydrogen production could, for example, be one transformation option relevant for seasonal energy storage, along with heat pumps and heat storage underground in basins, or in salts, depending on technology development and locational characteristics.

The 2DS and the resilient FSS

The 2DS has a reasoned mix of all options for energy supply with lowered emissions. It is an energy-only scenario in the sense that the details of a highly reduced, fossil energy use system are not modeled. How to fully adapt the housing stock to electricity and solar heat is left open, for example, as are the energy requirements for new infrastructure, and for other infrastructure requirements for a near-zero fossil fuel society. Such considerations will have played a background role in deciding what is reasonable during development of the 2DS model. Shifting from the reasonable 2DS to the resilient FFS eliminates options that cannot be ascertained in advance. The change in numbers relative to 2DS is summarized below; for overall results, see Table 1.3.

- The amount of remaining fossil energy is halved, due to a substantially lower amount of CCS.
- Biomass energy is reduced by a third, still doubling relative to 2015.
- Nuclear is calculated as constant relative to 2015, one third of 2DS.
- Hydro doubles relative to 2015, a quarter less than in 2DS.
- Geothermal rises sevenfold, but that is only half relative to 2DS.
- Other renewables also rise sevenfold, the same as in 2DS.
- Ocean energy rises 1,000-fold, from near zero, the same as in 2DS.

Total energy demand rises in line with economic growth of the OECD estimate of 2.6% and energy decoupling of 1% per year, as has been the case over the past 35 years.[10] Total primary energy supply then nearly doubles relative to the 2DS, from 669 exajoules (EJ) to 1,200 EJ. The gap in supply could be filled by solar

Table 1.3 The fail-safe scenario (FFS) compared with the 2-degree scenario for 2060

Current		*IEA 2-degrees scenario (2DS)*		*Fail-safe scenario (FSS)*	
2014		*2060*		*2060*	
182	Oil	99	Oil	45	Oil
165	Coal	46	Coal	23	Coal
119	Gas	90	Gas	45	Gas
54	Biomass	144	Biomass	100	Biomass
28	Nuclear	82	Nuclear	30	Nuclear
14	Hydro	40	Hydro	30	Hydro
2.7	Geothermal	44	Geothermal	20	Geothermal
1.4	Other renewables	9	Other renewables	4	Other renewables
0.004	Ocean	3	Ocean	3	Ocean
0.7	PV solar	31	PV solar	(...)	PV solar
0.1	Concentrated solar	44	Concentrated solar	(...)	Concentrated solar
0.1	Wind (offshore)	10	Wind (offshore)	(...)	Wind (offshore)
2.5	Wind (onshore)	27	Wind (onshore)	(...)	Wind (onshore)
569	**Total energy supply**	**669**	**Total energy supply**	**1,200**	**Total energy supply**
3.4	**Solar & Wind total**	**112**	**Solar & Wind total**	**900**	**Solar & Wind total**
	S&W share 0.6%		**S&W share 17%**		**S&W share 75%**
Growth 2010–2016 ~20%		**Yearly growth rate 8%**		**Yearly growth rate 24%**	

Source: http://www.iea.org/etp/explore/

Note
S&W, solar and wind; PV, photovoltaic.

and wind; the amount required to meet demand then rises from 112 EJ to 900 EJ, with the share of solar and wind rising from less than 0.6% now to around 75% in total primary energy supply. The sum of wind and solar would have to rise by 14% annually in the FSS. For comparison, the yearly growth rate from 2010 to 2016 was 16% for wind, 28% for solar, and 20% combined. Of course, at larger volumes the growth rates will tend to decrease (Kramer & Haigh, 2009), but incentives for faster expansion may well be increased. This near-full changeover to solar and wind renewables is not a final task for climate policy, but a fallback option to be reckoned with today. Instrumentations should also, in principle, be able to deliver in this worst-case fail-safe option. Primary energy in the FSS is transformed into final use energy with fewer losses than with fossil energy, depending on how much storage will be required, in turn depending on development in electricity markets, especially concerning decentral storage. Storage requirements are reduced with larger scale variable-priced markets. In the FSS, there would be abundant energy, with low emissions, if the remaining central use of fossils could be linked to CCS.

The climate policy task ahead: not just clean energy supply

The 2-degree goal, and certainly the 1.5-degree goal, requires a nearly complete transformation of all main sectors in society in order to wean them from fossil fuels. This is required even if other developments, such as population growth,

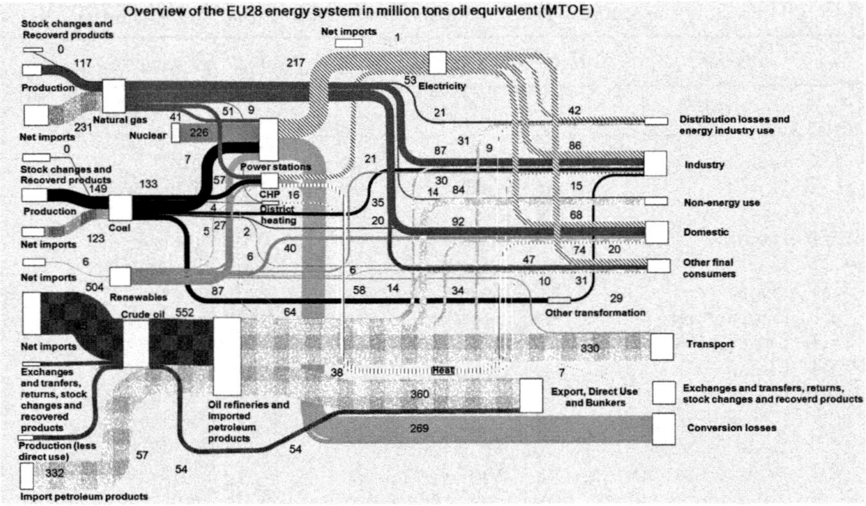

Figure 1.5 Energy input, transformation and final use in the EU-28.

Source: EEA (2016)[11]

reduced economic growth, vegetarian food choices becoming dominant, reduced transport volumes, and other not so probable or desirable developments, achieve lower emissions. Fossils currently fully dominate. Figure 1.5 depicts the energy flows of the 28 member nations of the EU. A similar, but interactive, Sankey flow chart is available for the world (see http://www.iea.org/etp/explore/). To ensure a livable future, all of the current main inflows on the left side of the figure will have to nearly disappear, replaced by renewables and some nuclear, with electricity and heat becoming the main types of energy use in final consumption, and a possible adjoining role for other energy carriers such as hydrogen, ammonia, aluminum, and others.

Planning & Control and Institutionalist climate policy instrumentations

The two main approaches to climate policy instrumentation have diverging starting points in political philosophies that have been present since ancient times. One seeks to decide centrally on what is best for all, that is, for the community. The other seeks to protect members of society against centralized public and private decisions as much as possible, with autonomy as a key to welfare. With the advent of democratic constitutions, protection against despotism has increased; in addition, what is best for all can be specified in ever greater detail, expanding domains of central state power. Private actions and the collective good can now be linked in two very different ways: by specifying the collective good and implementing what is needed to reach it, or by creating institutions under which decentralized action can flourish, guided for the good of all. In terms of climate policy, very different instrumentations will develop from each branch of political philosophy: a Planning & Control version and an Institutionalist version. They relate to the same society and so have intricate relations. In some domains, central planning and control must be dominant – for example, in military affairs when defending a country against foreign powers, or in coastal defenses against the seas, where institutions have a limited role to play. For most other domains, however, the choice is more open. Such is the case in climate policy and its instrumentation.

In developing climate policy instrumentation, the starting point in political philosophy must be clear, as it guides its strategic development. The starting point must be elaborated and linked to broader views on governance, the subject of Chapter 2. But such views must also relate to how society is functioning and may change its functioning, in political, economic-technical and cultural senses. The future must be discovered and it must be mastered, reckoning with what is more or less given and with what can be changed for the purpose of climate stabilization. The endpoint in reasoning is effectiveness: the reduction of emissions as required to keep global temperature rise this century *well below 2 degrees Celsius*, as stated in the 2015 Paris Agreement. In between is the instrumentation that will be needed to create a low-emission society by changing the current course of exploding emissions.

The reasoning involved in Planning & Control begins by identifying the technologies required to do the job, which form the core of policy development. The next point of focus is on the instruments capable of forcing the desired development, including the required administrative and organizational capacities. The political steps needed to decide, implement and maintain what is required follow. Finally, a limited number of infrastructure and institutional adaptations are required. In this top-down approach, the most relevant technologies must be selected, developed and brought into the market of our market-based economies. Choosing the right technologies and promoting them, creating the dynamics necessary to move toward the required low-emission developments and reducing the emitting technologies are the prime tasks of the Planning & Control approach. As technologies are substantially unknown over the long term, continuous planning plays a key role in order to adapt policy to what is realistically likely to arise next. This approach may be in line with current practical planning approaches to climate policy but should not be confused with this as yet ineffective climate policy instrumentation. New, more effective design must arise in the Planning & Control framework. That is the subject of Chapter 4.

Institutionalist reasoning begins by identifying what has become inadequate in current institutional arrangements and in public infrastructure. That outcome is the starting point for making repairs. Current institutions create incentives to emit ever-increasing amounts of greenhouse gases, because markets do not reckon with the external effects of climate emissions due *inter alia* to nonapplicability of liability rules in this domain. The main options for emission reduction are blocked by other inadequate institutions, such as those regulating electricity markets, with private monopolization in the background as a political development and problem. Improving institutions to deal with these new circumstances is the prime entry point for climate policy instrumentation. The goal of these changed institutions is to create the right incentives and options centrally, followed by public supply of infrastructure as required (for example, if this tends to be monopolistic, such as grid systems as for electricity). The Institutionalist view gives lower level administrative units and the private sector freedom to make decisions about production and consumption, as is usual for most economic decisions, and it envisions an open role for civil society. Bottom-up actions are the core result of climate policy, not dictated or guided by content from above.

The incentives and options created should together be able to reach the climate goal. This bottom-up Institutionalist approach is intended to induce the development and implementation of technologies not yet known. But how certain can we be that this approach will succeed? What bottlenecks will need to be resolved and options created, going beyond incentives, such as through infrastructure development required for future technologies? It seems clear already that a real-time, variable-price electricity market institution and long-distance, high-voltage direct current grids for electricity transport will be necessary. What more might be necessary? Such questions require a rough idea only of the possible useful technologies that may develop to allow for their bottom-up emergence and to prevent obstruction of potentially useful technologies. New institutions and infrastructure are developed in the Institutionalist framework. That is the subject of Chapter 5.

Reading guide

The two main instrumentation designs that are the centerpiece of this book are explored in detail in Chapters 4 and 5, following chapters that are included to guide the design process. The first of these, Chapter 2, focusing on political philosophy, is constitutive for instrumentation design, indicating basic choices in the design process. A more detailed method for instrumentation design is provided in Chapter 3, which distinguishes the primary directions for governance and links them to more general design considerations, such as types of instruments and how to arrive at a coherent overall design of effective instrumentations. A stepwise procedure is formulated for design in each of the two main governance directions. Specific key technology domains are approached from both governance directions, in order to illustrate the differences resulting from following one or the other governance direction. Next, in Chapters 4 and 5, a stepwise procedure is applied to instrumentation design in the two governance directions: Planning & Control and Institutionalism, respectively. Finally, Chapter 6 presents a comparison of the outcomes, referring back to the considerations developed in the first three chapters. A key issue discussed is the resilience of instrumentations to diverse unfavorable developments, such as political vulnerability in unfavorable economic periods accompanied by international tensions. Readers may wish to skip the more analytical Chapters 2 and 3 initially, going directly to the instrumentation designs of Chapters 4 and 5, and then return to these earlier chapters with more specific questions in mind later.

Notes

1. Source: http://www.esrl.noaa.gov/gmd/webdata/ccgg/trends/co2_data_mlo.png; retrieved April 2018.
2. See http://www.iea.org/publications/ and http://www.irena.org/home/index.aspx?PriMenuID=12&mnu=Pri#.
3. Economic data based on Maddison Project Data Base retrieved from http://www.ggdc.net/maddison/maddison-project/data.htm.
4. See http://www.wri.org/blog/2014/05/history-carbon-dioxide-emissions.
5. Source: https://data.giss.nasa.gov/gistemp/graphs/.
6. After the Smoot-Hawley Tariff Act (1930), U.S. imports were reduced by two thirds, hurting resource-poor countries most.
7. See https://data.worldbank.org/indicator/EG.USE.COMM.GD.PP.KD?view=chart.
8. Source: https://www.umweltbundesamt.de/sites/default/files/medien/378/publikationen/data_on_the_environment_2015.pdf; retrieved November 26, 2017.
9. Source: http://www.bloomberg.com/news/articles/2016-06-13/we-ve-almost-reached-peak-fossil-fuels-for-electricity.
10. For OECD prospective growth, see https://data.oecd.org/gdp/gdp-long-term-forecast.htm 23-06-2017. For decoupling empirically 1980–2014: -1.05%, see http://www.tsp-data-portal.org/Energy-Intensity-of-GDP#tspQvChart 23-06-2017.
11. Courtesy of Jeroen Peters, PBL Netherlands Environmental Assessment Agency, who edited the EEA figure for this book.

References

Acemoglu, D., & Robinson, J. A. (2012). *Why nations fail: the origins of power, prosperity and poverty*. New York: Crown Publishers.

Andersen, M. S., & Ekins, P. (2009). *Carbon-energy taxation: lessons from Europe*. Oxford, England: Oxford University Press.

Anderson, K., & Peters, G. (2016). The trouble with negative emissions. *Science*, *354*(6309), 182–183.

Arrhenius, S., & Holden, E. S. (1897). On the influence of carbonic acid in the air upon the temperature of the earth. *Publications of the Astronomical Society of the Pacific*, 9(54), 14–24.

Bauer, N., McGlade, C., Hilaire, J., & Ekins, P. (2018). Divestment prevails over the green paradox when anticipating strong future climate policies. *Nature Climate Change*, 8(2), 130–134. doi: 10.1038/s41558-017-0053-1

Boden, T. A., Marland, G., & Andres, R. J. (2017). *Global, regional, and national fossil-fuel CO2 emissions*. Oak Ridge, TN: Carbon Dioxide Information Analysis Center.

Brynjolfsson, E., & McAfee, A. (2014). *The second machine age: work, progress, and prosperity in a time of brilliant technologies*. New York, NY: Norton.

Buchanan, R. (1992). Wicked problems in design thinking. *Design Issues*, 8(2), 5–21.

De Koning, A., Huppes, G., Deetman, S., & Tukker, A. (2016). Scenarios for a 2^{0}C world: a trade-linked input–output model with high sector detail. *Climate Policy*, *16*(3), 301–317.

De Koning, A., Kleijn, R., Huppes, G., Sprecher, B., van Engelen, G., & Tukker, A. (2018). Metal supply constraints for a low-carbon economy? *Resources, Conservation and Recycling*, *129*, 202–208.

Deetman, S., Hof, A. F., & Van Vuuren, D. P. (2015). Deep CO2 emission reductions in a global bottom-up model approach. *Climate Policy*, *15*(2), 253–271. doi: 10.1080/14693062.2014.912980

Delbeke, J. (1992). Hacia un reforzamiento de la política medioambiental de la CE en los años 90. (English original: Towards a reinforced environmental policy for the 1990s.). *Revista de Economía Institucional*, *14*, 52–57.

Duchin, F., Levine, S. H., & Strømman, A. H. (2016). Combining multiregional input-output analysis with a world trade model for evaluating scenarios for sustainable use of global resources, Part I: conceptual framework. *Journal of Industrial Ecology*, *20*(4), 775–782.

EEA. (2016). Overview of the European energy system (ENER 036) assessment version, September 2015. Retrieved from http://www.eea.europa.eu/data-and-maps/figures/summaries-the-overall-picture-of-2

Elshkaki, A., Graedel, T. E., Ciacci, L., & Reck, B. K. (2016). Copper demand, supply, and associated energy use to 2050. *Global Environmental Change*, *39*, 305–315. doi: 10.1016/j.gloenvcha.2016.06.006

EU. (2012). *Energy efficiency directive 2012/27/EU of the European Parliament and of the Council*. Luxembourg: Official Journal of the European Union.

Flohn, H. (1977). Climate and energy: A scenario to a 21st century problem. *Climatic Change*, *1*(1), 5–20.

Font Vivanco, D., Freire-González, J., Kemp, R., & Van der Voet, E. (2014). The remarkable environmental rebound effect of electric cars: a microeconomic approach. *Environmental Science & Technology*, *48*(20), 12063–12072.

Font Vivanco, D., Kemp, R., & van der Voet, E. (2015). The relativity of eco-innovation: environmental rebound effects from past transport innovations in Europe. *Journal of Cleaner Production*(0). doi: 10.1016/j.jclepro.2015.04.019

Font Vivanco, D., & van der Voet, E. (2014). The rebound effect through industrial ecology's eyes: a review of LCA-based studies. *The International Journal of Life Cycle Assessment*, *19*(12), 1933–1947.

Goodwin, P. (2012). *Peak travel, peak car and the future of mobility. Evidence, unresolved issues, policy implications, and a research agenda*. Paper presented at the Roundtable on Long-Run Trends in Travel Demand, November 29–30.

Graedel, T. E., & van der Voet, E. (2010). *Linkages of sustainability*. Cambridge, MA: MIT Press.

Grubb, M., Bashmakov, I., Drummond, P., Myshak, A., Hughes, N., Biancardi, A., ... Lowe, R. (2018). *An exploration of energy cost, ranges, limits and adjustment process. Final report to the Institute for New Economic Thinking*. Retrieved: https://www.ucl.ac.uk/bartlett/sustainable/sites/bartlett/files/an_exploration_of_energy_cost_ranges_limits_and_adjustment_process_2_0.pdf

Heck, V., Gerten, D., Lucht, W., & Popp, A. (2018). Biomass-based negative emissions difficult to reconcile with planetary boundaries. *Nature Climate Change*, 1.

Hertwich, E. G., Gibon, T., Bouman, E. A., Arvesen, A., Suh, S., Heath, G. A., ... Shi, L. (2015). Integrated life-cycle assessment of electricity-supply scenarios confirms global environmental benefit of low-carbon technologies. *Proceedings of the National Academy of Sciences*, *112*(20), 6277–6282.

Hovi, J., Sprinz, D. F., & Bang, G. (2012). Why the United States did not become a party to the Kyoto Protocol: German, Norwegian, and US perspectives. *European Journal of International Relations*, *18*(1), 129–150. doi: 10.1177/1354066110380964

Huppes, G., & Ishikawa, M. (2009). Eco-efficiency guiding micro-level actions towards sustainability: ten basic steps for analysis. *Ecological Economics*, *68*(6), 1687–1700.

ICSG. (2017). *The world copper factbook 2017*. Lisbon: ICSG Retrieved from: http://www.icsg.org/index.php/component/jdownloads/finish/170/2462

IEA. (2011). *25 Energy efficiency policy recommendations, 2011 update*. Paris: International Energy Agency.

IEA. (2016). *World Energy Outlook 2016*. Paris: International Energy Agency.

IEA. (2017a). *Energy efficiency indicators: Highlights 2016*. Paris: IEA.

IEA. (2017b). *Energy technology perspectives 2017, catalysing energy technology transformations*. Paris: IEA

Interpol. (2013). *Guide to carbon trading crime*. Lyon, France: International Criminal Police Organisation (Interpol).

IPCC. (2014). Summary for policymakers of WGIII. In: O. Edenhofer, R. Pichs-Madruga, E. Sokona, Y. Farahani, S. Kadner, K. Seyboth, … J. C. Minx (Eds.), *Climate change 2014, mitigation of climate change. Contribution of Working Group III to the Fifth Assessment Report of the Intergovernmental Panel on Climate Change*. Cambridge: Cambridge University Press.

IPCC-AR5-WGIII. (2014). *Mitigation of climate change. Contribution of Working Group III to the Fifth Assessment Report of the Intergovernmental Panel on Climate Change*. Cambridge: IPCC. Retrieved from https://www.ipcc.ch/pdf/assessment-report/ar5/wg3/ipcc_wg3_ar5_frontmatter.pdf

IPCC-WG3. (2014). *Working Group III – Mitigation of climate change, annex III: technology-specific cost and performance parameters*. Retrieved from https://www.ipcc.ch/pdf/assessment-report/ar5/wg3/ipcc_wg3_ar5_annex-iii.pdf

Jackson, T. (2011). *Prosperity without growth: economics for a finite planet*. New York: Routledge.

Jensen, S., Mohlin, K., Pittel, K., & Sterner, T. (2015). An introduction to the Green Paradox: The unintended consequences of climate policies. *Review of Environmental Economics and Policy*, 9(2), 246–265.

Jordan, A., Rayner, T., Schroeder, H., Adger, N., Anderson, K., Bows, A., … Whitmarsh, L. (2013). Going beyond two degrees? The risks and opportunities of alternative options. *Climate Policy*, *13*(6), 751–769. doi: 10.1080/14693062.2013.835705

Kallis, G. (2011). In defence of degrowth. *Ecological Economics*, *70*(5), 873–880.

Kesicki, F., & Ekins, P. (2012). Marginal abatement cost curves: a call for caution. *Climate Policy*, *12*(2), 219–236. doi: 10.1080/14693062.2011.582347

Kramer, G. J. (2017). *De ontdekking van de toekomst (The discovery of the future)*. Utrecht, Netherlands: Utrecht University.

Kramer, G. J., & Haigh, M. (2009). No quick switch to low-carbon energy. *Nature*, *462*(7273), 568–569.

Lackner, K. S., & al, e. (2016). Letter on the promise of negative emissions. *Science*, *354*(6313), 714–714. doi: 10.1126/science.aal2432

Levin, K., Cashore, B., Bernstein, S., & Auld, G. (2012). Overcoming the tragedy of super wicked problems: constraining our future selves to ameliorate global climate change. *Policy Sciences*, *45*(2), 123–152.

Mazzucato, M. (2013). *The entrepreneurial state: debunking public vs private sector myths* (vol. *1*). London: Anthem Press.

McGlade, C., & Ekins, P. (2015). The geographical distribution of fossil fuels unused when limiting global warming to 2 degrees C. *Nature*, *517*(7533), 187–190.

Mokyr, J. (2004). *The gifts of Athena: historical origins of the knowledge economy*. Princeton, NJ: Princeton University Press.

Moura, M. C. P., Smith, S. J., & Belzer, D. B. (2015). 120 Years of U.S. residential housing stock and floor space. *PLoS ONE*, *10*(8), e0134135. doi: 10.1371/journal.pone.0134135

Nakamura, S., Nakajima, K., Kondo, Y., & Nagasaka, T. (2007). The waste input- output approach to materials flow analysis. *Journal of Industrial Ecology*, *11*(4), 50–63.

Nakicenovic, N., Lempert, R. J., & Janetos, A. C. (2014). A framework for the development of new socio-economic scenarios for climate change research: introductory essay. *Climatic Change, 122*(3), 351–361. doi: 10.1007/s10584-013-0982-2

O'Neill, B. C., Kriegler, E., Ebi, K. L., Kemp-Benedict, E., Riahi, K., Rothman, D. S., . . . Solecki, W. (2015). The roads ahead: narratives for shared socioeconomic pathways describing world futures in the 21st century. *Global Environmental Change*. doi: 10.1016/j.gloenvcha.2015.01.004

OECD. (1992). *Climate change. Designing a practical tax system*. Paris: Organisation for Economic Co-operation and Development.

Olivier, J. G. J., Janssens-Maenhout, G., Muntean, M., & Peters, J. A. H. W. (2014). *Trends in global CO2 emissions: 2014 Report*. The Hague: Planbureau voor de Leefomgeving. Retrieved from http://www.pbl.nl/en/publications/trends-in-global-co2-emissions-2014-report

Pages. (2016). Interglacials of the last 800,000 years, Past Interglacials Working Group of PAGES. *Reviews of Geophysics*, *54*(1), 162–219. doi: 10.1002/2015RG000482

Piketty, T. (2014). *Capital in the twenty-first century*. Cambridge, MA: Belknap Press of Harvard University.

Pimentel, D., & Pimentel, M. H. (2007). *Food, energy, and society* (3rd ed.). Boca Raton. FL: CRC Press.

Ramos, F. M., Bambace, L. A. W., Lima, I. B. T., Rosa, R. R., Mazzi, E. A., & Fearnside, P. M. (2009). Methane stocks in tropical hydropower reservoirs as a potential energy source. *Climatic Change*, *93*(1), 1. doi: 10.1007/s10584-008-9542-6

Raworth, K. (2017). *Doughnut economics: seven ways to think like a 21st-century economist*. White River Junction, VT: Chelsea Green Publishing.

Riahi, K., Van Vuuren, D. P., Kriegler, E., Edmonds, J., O'Neill, B. C., Fujimori, S., . . . Fricko, O. (2017). The shared socioeconomic pathways and their energy, land use, and greenhouse gas emissions implications: an overview. *Global Environmental Change*, *42*, 153–168.

Rifkin, J. (2014). *The zero marginal cost society. The internet of things, the collaborative commons, and the eclipse of capitalism*. New York: Palgrave Macmillan.

Rittel, H. W., & Webber, M. M. (1973). Dilemmas in a general theory of planning. *Policy Sciences*, *4*(2), 155–169.

Rogelj, J., Schaeffer, M., Friedlingstein, P., Gillett, N. P., van Vuuren, D. P., Riahi, K., . . . Knutti, R. (2016). Differences between carbon budget estimates unravelled. *Nature Climate Change*, *6*(3), 245–252. doi: 10.1038/nclimate2868

Samuelson, P. A. (1955). Diagrammatic exposition of a theory of public expenditure. *Review of Economics and Statistics*, 350–356.

Saunders, H. D. (2015). Recent evidence for large rebound: elucidating the drivers and their implications for climate change models. *Energy Journal*, *36*(1), 23–48.

Schaller, M. F., Fung, M. K., Wright, J. D., Katz, M. E., & Kent, D. V. (2016). Impact ejecta at the Paleocene-Eocene boundary. *Science*, *354*(6309), 225–229. doi: 10.1126/science.aaf5466

Scheidel, W. (2017). *The great leveler: violence and the history of inequality from the Stone Age to the twenty-first century*. Princeton. NJ: Princeton University Press.

Schipper, B. W., Lin, H.-C., Meloni, M. A., Wansleeben, K., Heijungs, R., & van der Voet, E. (2018). Estimating global copper demand until 2100 with regression and stock dynamics. *Resources, Conservation and Recycling*, *132*, 28–36.

Schumpeter, J. A. (1947). *Capitalism, socialism, and democracy*. New York: Harper & Brothers. (Originally published 1942.)

Solano, B., & Drummond, P. (2014). Techno-economic scenarios for reaching Europe's long-term climate targets using the European TIMES model (ETM-UCL) to model energy system development in the EU. In CECILIA2050 (Ed.), *Choosing efficient combinations of policy instruments for low-carbon development and innovation to achieve Europe's 2050 climate targets*. London: University College London (UCL).

Sorrell, S. (2009). Jevons' paradox revisited: the evidence for backfire from improved energy efficiency. *Energy Policy*, *37*(4), 1456–1469. doi: 10.1016/j.enpol.2008.12.003

Stephenson, A., & MacKay, D. J. C. (2014). *Life cycle impacts of biomass electricity in 2020. Scenarios for assessing the greenhouse gas impacts and energy input requirements of using North American woody biomass for electricity generation in the UK*. (URN 14D/243). London: Department of Energy & Climate Change (DECC). Retrieved from http://www.gov.decc.

Suh, S. (2009). *Handbook of input-output economics in industrial ecology* (vol. 23). New York: Springer Science & Business Media.

Sunstein, C. R. (2014). *Simpler: the future of government*. New York: Simon and Schuster.

UBA. (2015). *Data on the environment 2015*. Berlin: UmweltBundesAmt.

Van Bavel, B. (2016). *The invisible hand? How market economies have emerged and declined Since AD 500*. Oxford, UK: Oxford University Press.

Weitzman, M. L. (2009). On modeling and interpreting the economics of catastrophic climate change. *Review of Economics and Statistics*, *91*(1), 1–19.

Weitzman, M. L. (2011). Fat-tailed uncertainty in the economics of catastrophic climate change. *Review of Environmental Economics and Policy*, *5*(2), 275–292. doi: 10.1093/reep/rer006

Wells, H. G. (1913). *The discovery of the future*. New York: BW Huebsch. (Originally published 1902.)

Zellou, A. M., & Cuddington, J. T. (2012). *Trends and super cycles in crude oil and coal prices*. Golden, CO: Colorado School of Mines.

2 Political philosophy for climate policy instrumentation

Political philosophy determining governance for climate policy

Choices in democratic liberalism and beyond

Climate policy seems like a logical task for governments as a means of solving the emerging climate problem. It is part of creating the good for all, because damage caused by climate change is much larger than the costs of its prevention. This reasonable position, which follows from the collective welfare view that rose to prominence in the 18th century, states that one role of government is to create the greatest happiness for the greatest numbers of people (Bentham, 1776). This view deviates from earlier views of absolutist government, in which the sovereign or a central leadership decided what was best for the country, not necessarily for the collectivity of its inhabitants (Bodin, 1576/1992; Hobbes, 1651/1968).

The American and French revolutions brought these philosophical frameworks together. Transforming the political system into a democracy can lead central government to do what is best for all, in any relevant domain. In a nutshell, this is democratic welfare liberalism, which has become operational and is now mostly dominant in political parties and governments as a new version of absolutism. Key concepts in 20th-century liberal democratic development are operational decision criteria, as in Pareto optimality and the potential Pareto improvement, whereby winners might compensate losers fully (Schumpeter, 1949; Stavins, Wagner, & Wagner, 2003); collective decision theory, identifying some problems but leaving the core democratic principles intact (Arrow, 1951/2012; Maskin & Sen, 2014); and operational welfare theory (Samuelson, 1977), expanding to all domains of human welfare and development (Sen, 1970; Nussbaum & Sen, 1993; Stiglitz, Sen, & Fitoussi, 2009). Within this broad democratic welfare domain there are substantial differences along the right–left dimension but also, more broadly, across economic, social and moral perspectives, ranging from communism and state capitalism, to social democracy and liberalism, to conservatism and neo-liberalism. Political views of how to approach that welfare ideal are deeply divided. What unites these disparate approaches is that governments have a responsibility for collective welfare. Long-term climate policy should not tread on the various and changing views of diverse political parties, bodies and persons but rather should rest on some overarching central view or principle. Going back to such basics might well involve returning to the classic philosophers Plato and Aristotle to explore the evolution of these diverging views.

In Plato's governance view, expounded in the *Republic* (c. 380 BCE) and the Laws, see Bobonich and Meadows (2018) for a comparative analysis.[1] The state is a separate entity with a responsibility for the whole, even if this entails going against the interests of specific groups and persons, including the sacrifice of individual lives for the common good (Brown, 2011). From this perspective, the death sentence imposed on Socrates might be justified. The ethical and normative requirements on rulers and ruled are expected to provide safeguards against misuse, not a restriction in domains of state power. Although Plato advocates for an ideal state presided over by wise philosopher-kings, Popper places his political views close to those of authoritarian if not fascist regimes (Popper, 1945/1966/2012), seemingly in contradiction to the option of a wise king deciding for others what is best. In the Aristotelian perspective, there certainly are domains where the state is necessary, creating preconditions. But in most domains individuals differ and know what is best for themselves, in their individual ways. Constitutional arrangements focus on a number of necessarily public provisions such as military defense and public infrastructure and conflict resolution. But they have a main focus on preventing misuse and usurpation of power (Aristotle, 330–322 BC, 1950).[2] The constitution as the core institution allows for adjustments in all other institutions, with safeguards on misuse to be always maintained. Both approaches may be seen as seeking the common good. Modern democratic welfare states can be viewed along the continuum of these two lines, as specifying the common good in detail and doing what is required in detail, or as creating the institutions allowing for decentralized organizations and decisions for the good of each, within the common framework of the state. They comprise the philosophy of liberalism, taken broadly.

Effective climate policy must be implemented based on political support, with science, the press, political parties and more playing a role. That support, in turn, is grounded in political philosophy. Almost all political parties link to both philosophical branches, with varying emphasis in varying domains, making support for either perspective in climate policy instrumentation a matter of discussion and choice. Some political movements go beyond these two main branches of political philosophy; for example, libertarians may deny any broader role of the state beyond national safety, leaving the domain of democratic optimality for the common good.

Institutions-oriented political philosophy for long-term economic development

The two main branches of Western political philosophy balance central effectiveness with a short-term focus against autonomy of decentralized organizations with long-term consequences. From an Institutionalist perspective, Western industrialization was made possible by the decentralization of political power that occurred after the collapse of the Roman Empire. The new, weaker states that remained allowed for a first round of medieval technology development in the latter part of the Middle Ages, with a key role for hydropower (Gimpel, 1977). New momentum with full-fledged economic growth accompanied the technology revolution of the late 16th century, and expanded into the 17th and 18th centuries with the use of wind, water, peat and coal (Warde, 2007). Coal became dominant in the 18th century, with science and metallurgy leading to the development of the steam engine. The Institutionalist dominance continued until late in the second half of the 20th century, alternating with return to centralized absolutist government in the eras of Philip II of Spain, France's Louis XIV, Napoleon, and, most recently, Hitler and Stalin. In each historical period, centralized states eventually lost out economically – as well as politically – to more loosely organized, more inclusive and more democratic countries, which were also able to mobilize the power and creativity of large parts of the population when needed, such as in times of war. Following victory in World War II (WWII), the Conservative government led by Churchill lost the elections of 1945 as the political winds shifted, with voters viewing the opposing Labour Party as best suited to develop Britain economically and socially.

Core institutions for long-term Western development were a substantial separation of the organizations of church and state; substantially independent universities; a restricted domain of central government, which for kings was limited to maintaining public order and the military only; a private law domain for trade and conflict resolution, which was derived from Roman law; and a great deal of decentralized power in cities and other local governmental bodies. The diverse sources of power supported development of the rule of law as a means of resolving conflict on the basis of justice and established precedent, not power. Burgeoning international trade in the Hanseatic League, a pillar of modern economic and organizational development that preceded the founding of the World Trade Organization by six centuries, was arranged among substantially independent cities of different weak states, and collapsed when it was centralized as a monopoly under control of the city of Lübeck.

Institutionalism still dominates post-WWII, with unprecedented growth

After WWII, the new economic institutions for business cycle stabilization, monetary stability and international trade unleashed a period of global economic growth never before seen in history. At the same time, aspects of the two formerly competing views of government merged. Concepts such as centralization for the common good were combined with democratic safeguards against power usurpation, based on effective universal suffrage and constitutional restrictions. Restraints on central power lost their attractiveness as democratic governments were viewed as being in the best position to know what is good for all, aided by newly developed welfare theory, which became increasingly operational with the rise of economic science and information technology. In these governments, the core idea of democratic collective welfare guides public action: governments are not merely permitted to do what is best for all, but are expected to do so for the common good. In addition to economic welfare, all other aspects of life contributing to the common good became part of central public policy, with great success. Economic growth eliminated hunger and reduced poverty; education was democratized; all were invited to participate in cultural activities for rich self-development; and the logic of welfare economics dictated that women be emancipated and that all minorities be included on an equal basis with members of formerly dominant groups. Institutionalism was losing ground to more detailed measures and the detailed policy instruments they required.

Unity in political philosophy begins to erode

This new unity in political philosophy lasted for most of the last half century. But cracks formed from within and without as other strands of political philosophy clamored for attention. The core of democratic welfare liberalism became stressed internally because of an overload of required information and the diversity of worthwhile goals. These goals no longer provide clear guidance on public policy, as was vividly shown in the UN's 17 Sustainable Development Goals (SDGs), which include 230 operational indicators setting more specific targets that are to be further detailed according to a number of additional criteria. Specific recommendations abound, but the welfare integration of solutions has become virtually impossible. Democratic welfare liberalism is also under stress due to the growing political presence of competing political philosophies, especially extreme libertarianism and a more autocratic, conservative version of absolutism. Political stability and legitimacy have decreased. It is noteworthy that the 10% of the global population not experiencing a rise in primary income in the past quarter century happens to include the poorer 50% of developed countries, equal to roughly 50% of all voters in those countries (Milanovic, 2016); these trends are shown in Figure 2.1.

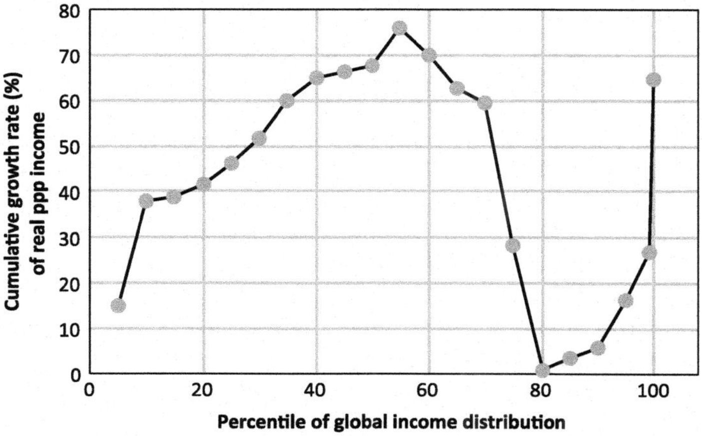

Figure 2.1 Global growth in primary income, 1988–2011: the elephant curve.

Source of data: Milanovic (2016)

Climate policy remains to be developed amid the philosophical turmoil

Climate policy must deal with such broad developments in order to be effective, and it must be resistant to other distracting philosophical and political developments. The two essentially different approaches to liberal governance have existed for millennia and underlie the still dominant domain of democratic welfare liberalism. However, the merger of these philosophical approaches in the years after WWII was counterproductive for climate policy, and a clear separation may need to be reestablished. One approach has a clear focus on adequate central government powers for creating the collective good and would seek far-reaching improvement of current technology-specific climate policy instrumentation. The other promotes decentralized autonomy within much more abstract and general institutions and creates incentives and options for emissions reduction. Autonomy safeguarded by constitutional arrangement detracts from direct, centrally steered effectiveness, but may be more effective in the long term. These two branches within liberal political philosophy are linked to forms of governance referred to as Planning & Control and Institutionalist governance – alternatives that are also applicable in the field of climate policy. They lead to two fundamentally different and mutually exclusive sets of climate policy instrumentation, which are the focus of this book.

Modern political philosophies and climate policy

Versions of liberalism dominate

The currently dominant Western political philosophy has a liberal orientation, with many shades of grey, ranging from social democratic–liberal to more conservative and neo-liberal. All such variants of liberalism link to different modes of governance ranging from the development of concrete policy instruments and actions to more institutional orientations. These main variants must all deal with new developments in technology, society and the environment – in this case, the climate issue. Current political debates within the political center relate to applied democratic welfare optimization, with different views on how to define and move toward the never fully reachable optimum. The action-oriented variant of liberalism has now become dominant nationally, internationally, and in global organizations such as the United Nations Environment Programme (UNEP), and the two UN organizations specifically concerned with the climate – the United Nations Framework Convention on Climate Change (UNFCCC) and the Intergovernmental Panel on Climate Change (IPCC). The other, more general, institutional line in Western liberalism has guided institutional development over the past centuries, leading to the Industrial Revolution and more recently to two periods of unprecedented global economic growth. Today's climate problem is a result of that growth explosion after 1960. The solution to the climate problem will most likely be sought within these two dominant strands in liberalism, yielding two substantially differing options for instrumentation. The centralized planning approach now constitutes the main entry into climate policy. But that planning approach is still mixed, including incompatible institutionalist elements, and will need to become much more stringent and encompassing in order to be effective. The institutionalist approach, as reflected in the OECD "polluter pays principle," must be newly developed for climate policy instrumentation. Both will have to operate under new, not-so-liberal political pressures, and may require broader actions that go beyond climate policy, such as countering the tendency toward economic and political monopolization and related inequality.

Other strands of political philosophy

Non-liberal strands of political philosophy are present in the Western world and in non-Western regions, as well. Some extreme examples demonstrate how they may relate to climate policy and its instrumentation. Libertarians give a dominant place to freedom, expressed in two different ways. The anarcho-capitalist version of the political philosophy of libertarianism is primarily concerned with protection of the individual from state power. It promotes a simple constitution with narrowly defined public tasks and the avoidance of detailed regulation. Especially in the United States, this branch of philosophy has powerful followers and intellectual leaders. Examples are Murray Newton Rothbard (Hoppe, 1999; Rothbard, 2009) and Ayn Rand, both inspired by the Austrian School of political economics of Ludwig von Mises. In this line of political philosophy, public climate policy-making is to be avoided as not being a concern of the state. The climate problem, not denied, is to be resolved, if at all, by private action to create and supply low-emission technologies, and by users who choose to acquire and use low-emission technologies. No specific climate policy instrumentation follows from this view, as the cure of active public policy is worse than the disease. Bad luck is always a possibility, and then the brightest minds are the best for dealing with challenging circumstances.

The second version of libertarianism protects individuals from both the powerful state and powerful capitalist monopolies. In this communal left-wing version of libertarianism, a branch of political philosophy now mostly present in Latin speaking countries, private ownership of the means of production is replaced by common or cooperative ownership that links economy and state, but only at a very local and decentralized level. Coordinated climate policy, again, seems a distant need, with many small and independent communities not easily collaborating even if they want to do so. But dismantling centralized economic institutions, including for the purpose of inducing economic growth, would cause emissions to decline by returning to an ideal of local autonomy with limited material welfare. The "degrowth" movement reflects this direction (D'Alisa, Demaria, & Kallis, 2014; Kallis, 2011).

At the boundary of liberalism, an Edmund Burke type or Bismarck type of conservativism would yield different results given that cultural identity, including the dominant religion of the country, would play a central role in creating unity. Opportunities for social and environmental policy would be limited, but an opening for climate policy could exist, based on principles of public peace, safety, and order. Economic strength is a prerequisite for national power. U.S. climate policy proposals presented by former Secretary of State James Baker are an example of how this brand of conservatism plays out (Baker et al., 2017). Protecting the country against climate damage then becomes a public task; however, permitting international action to infringe on national sovereignty is a route to be avoided. Binding international agreements are not acceptable. National climate policy should avoid prescribing direct regulations as much as possible, with instruments oriented toward institutional redesign but accepting large dominant firms in most domains as key actors for the promotion of national progress.

Significant parts of the world are governed on the basis of other and mostly mixed political philosophies, including China, India, Indonesia and Russia, and perhaps also Japan as a mainly single-party country. Substantial internal variation exists. But their governing philosophies are linked more or less to variants of Western political philosophy, including communism and socialism. Communist state-capitalism is the Chinese variant. Russia has turned its back on communism but reverted mainly to a state-capitalist base structure. It is not yet clear what direction India and Indonesia will take. In all of these countries the elites are extremely wealthy, with the numbers of billionaires rising sharply (UBS & PWC, 2017), closely linked to political power. The United States seems to be moving in the direction of combined economic and political power as well, more so than European countries. Such developments may lead to conservation of interests of the rich, and then to stagnation (Acemoglu & Robinson, 2012; Van Bavel, 2016). Climate policy, in that case, is not a priority.

In a cultural sense, democratic welfare liberalism is still dominant. All major countries intend to advance the common good, including the BRIC countries (Brazil, Russia, India and China). However, most non-Western countries have fewer constitutional safeguards than their Western counterparts, and economic and political inequality may similarly move them toward stagnation in the Acemoglu and Robinson manner. Similar macro-political reasoning, also including Western countries, can be found in the work of Scheidel (2017), Milanovic (2016), Van Bavel (2016). Piketty (2014), Smil (2009). Although tendencies and mechanisms are clear, outcomes are not seen as deterministic: there are other solutions before war and other disaster reduce inequality.

Political philosophies have been competing since ancient times

Central control versus autonomy

The philosophies of Plato and Aristotle, and subsequent approaches, offer varying perspectives on how to balance central control and autonomy, Plato accepted the death penalty imposed on Socrates by the rulers of Athens' institutions (Popper, 1945/1966/2012). Aristotle, sensing political trouble in Athens, escaped sentencing by fleeing to Macedonia. His line of reasoning concerning protection against the usurpation of central power was worked out in detail in his *Constitution of Athens*. The roles of the state and of private belief were already somewhat separate in early Christianity. Matthew 22 states, "repay to Caesar what belongs to Caesar and to God what belongs to God," indicating at least some separation of the two domains, and limiting the domain of the state. State and church may align, but should not coincide. The Roman Empire first did exactly that, with Hadrian unifying by force all regional religions in one framework, roughly coinciding in time with the end of technological and economic growth and political expansion of the Empire. However, the Romans left another substantial domain largely beyond direct central political control: the practice of private law (M. Smith, 1904). Emperor Justinian tried to unify private law in the *Codex Justinianus* (Body of Civil Law) but did not achieve much beyond stating which laws and sources of laws existed; central courts, for example, were not installed. Roman private law remained dominant in the Western world until after the WWII, accommodating private contracts and private trade. Agrarian feudal society in the Middle Ages had few constraints imposed on power. In parallel, however, there was considerable decentralized freedom in cities, with courts mostly local and independent of the central state, and completely new, substantially independent universities (Bloch, 1939/1961; Hilton, 1995). The late Middle Ages witnessed the first Western industrial revolution ranging over vast areas of Roman Catholic Europe (Gimpel, 1977).

Efficiency, democracy and inequality

Democratic welfare theory, in line with Bentham's "greatest happiness for the greatest numbers," did not become operational until after WWII. By the 1960s, the social-democratic version of liberalism had become dominant, in competition with more extreme neo-liberal versions of liberalism. Both place no constraints on domains of control, with neo-liberals shifting economic power to large private parties in their constituency. Democratic welfare theory is foundational to both, but they envision different routes to broadly defined economic development. The balancing act between centralized efficiency and decentralized autonomy tilted toward centralization in both versions of liberalism, as well as the so-called Third Way, a centrist position outlined by Giddens (1998). Primary income and wealth inequality began to rise again after 1970. Although linked to technological development and globalization, these changes were also evident in political philosophy, associated with the ideas of the political philosopher John Rawls (1971/2009). He argued that under an assumed *veil of ignorance*, any person would prefer societal arrangements that improve the lot of the most deprived, as the deprived person could very well be him- or herself. This premise holds true even when the much greater increase in income of the rich is implied, as long as this extreme rise in income is instrumental for improving the lot of the poorest. Using this argument, tax reductions for the rich could now be stated in democratic welfare terms as being "done for the poor." The empirical link to improvement of the poorest has not been substantiated, however; see the development of income inequality in

Figure 2.1. Redistribution of income and wealth by means of progressive taxes, old-age pensions, and low-cost public provision of health and education has compensated substantially for increasing inequality in most North-Western European countries but not in Anglo-Saxon countries.

Liberal central effectiveness versus liberal autonomy and incentives

Current climate policy for the most part uses technology-specific instrumentation. Results are meager at best, however, with concentrations of greenhouse gases in the atmosphere rising ever more rapidly, especially CO_2. Stronger action by means of central planning is one option for creating more effective climate policy. This requires new, more systematic and powerful instrumentation in order to effect climate goals. Reforming institutions is the other option. Institutions work less directly and over the longer term, but they may have a deeper influence, as shown in the centuries since the start of the Industrial Revolution. Their design has a more general foundation in legal, ethical and economic principles. Institutional rearrangement to resolve undesirable consequences requires a rethinking of the institutions that shaped Western development toward economic growth, now extending globally. The problems that Western and now global development have caused, from economic and political inequality to climate change, must be resolved by institutional innovation. That return to and reorientation of institutions can form a second approach to developing climate policy instrumentation.

The first governance approach – planning for the common good – spans the centuries from Plato's *Republic*, to Hobbes's *Leviathan*, to the welfare approach of Bentham's "greatest happiness for the greatest numbers," to Rousseau (1772/2008) and the French Revolution. A direct line can also be drawn to the operational welfare theory formulated after WWII (Samuelson, 1977; Sen, 1970), its broadening to include ethical and moral issues (Nussbaum & Sen, 1993), and to also reckon with common market imperfections, as outlined by Stiglitz et al. (2009). Democracy serves as a safeguard, ensuring that the best approach is, indeed, realized. All persons are represented in decisions, rather than relying on a wise sovereign. Direct democracy is the ideal, with representative democracy implemented for practical reasons, only. Applied welfare theory, as in cost-benefit analysis, is incorporated to smooth the detailed political decision-making process, obligatory in the United States since President Reagan's Executive Order (E.O.) 12291 of February 17, 1981, and with somewhat similar regulations in the EU.

The other governance approach for the common good has its focus on central institutions, creating decentralized incentives, and options and constraints, with mostly decentralized practical decisions regarding what seems best for the different parties involved, according to their own views. In industrialized societies, most economic decision making was decentralized within the central institutional framework developed during the last several centuries, and built on the continuity of Roman private law. At the core of this institutionalist approach is freedom from monopoly, both private and public. Developing oneself independently is central to personal welfare, always in a social and political context, under adequate economic circumstances. There is no benevolent central body in an autocratic sense, as in Hobbes's *Leviathan*, or in the democratic welfare ideal after the French Revolution. Although there are some domains of direct central action, the main action is indirect, through liability rules, court procedures and decentralized public bodies such as cities, water boards, harbor boards and more.

It is this decentralized autonomy within an overall generic institutional framework that allowed for and gave rise to the Industrial Revolution and later global economic growth. The steam engine was not the result of a public investment program, and its further development was halted for over a decade when James Watt managed, through political maneuvering, to be granted a patent-based monopoly. Watt spent his time on political intrigues and litigation against intruders on this newly created patent monopoly. Adam Smith, an adherent of the school of decentralized individual liberty, outlined this more institutional direction in his *Wealth of Nations (1776/1979)*. He advocated both an independent, private economic domain, with no direct interference (let alone control) in concrete economic decisions by government and, conversely, independent government immune to takeover by private firms and interests. Intermittent, absolutist deviations such as occurred in Spain during the reign of Philip II and in France, with Louis XIV, delayed the economic and social development of those nations. The essential characteristic in institutionalist governance is that there is not a single highest-order decision-maker representing society charged with doing the best for all. Democracy, in this scenario, is characterized by negotiation and discussion, limiting central power over decentralized decisions as much as possible, and safeguarding against all public and private usurpation of decision-making authority.

In current terms, technical decisions regarding the economy should not be implemented centrally even if decided democratically. If the amount of steel to be used in the construction of a bridge is decided democratically, it will most certainly collapse. If governments overrule central banks in their role as guardians of monetary stability, such stability will be reduced. Specialized groups must have only limited domains of authority, however, which cannot normally be overruled by centralized, general governmental bodies. Centralized authority involves only the most abstract rules, governing decision making in society indirectly and procedurally, with conflict resolution as the core. This view can be linked to diverse philosophers starting with Aristotle, and in Western societies with several centuries of Magna Carta discussions. Independent universities and substantially independent courts and regional governments all arose during the Middle Ages. There was not much central power regionally, and certainly not in cities. Cross-continental, horizontal linkages such as those organized by the Church of Rome led to the development and diffusion of new technologies and organizations. These, in turn, led to the first industrial revolution in Europe, as convincingly argued by Gimpel (1977). Ostrom won a Nobel prize for her groundbreaking research on the essential role of autonomous bottom-up processes on societal and economic development (Ostrom, 2010), reflecting the institutionalist countercurrent against central planning in the current, mostly centralizing, version of liberalism.

Political philosophies for climate policy

Climate governance choices must be viewed not only against the backdrop of current empirical social, economic and political developments, but also against the long-term historical backdrop of overarching political theory and governance developments that influence all policy decisions, including the choice of instrumentation for climate policy. As Table 2.1 highlights, political philosophy and

Table 2.1 Political views on centralist state power versus decentralized domains of power

Centralist state power in all domains / Monistic actions for the common good	*Institutionalist multiple domains of power / Incentives and options*
Plato's *Republic* (~380 BCE)	Aristotle's *Constitution of Athens* (330–322 BCE) Separation of state and religion: Matthew 22
Roman Empire: merger of state and uniform religion under Hadrian (±125 AD)	Roman Empire: state-independent private law (M. Smith, 1904)
End of Western Roman Empire: feudal society (Bloch 1939/1961; Hilton, 1995)	End of Western Roman Empire: independent courts and universities First industrial revolution (Gimpel, 1977)
State and religion merged: Lutheran and Calvinist Reformation, Catholic Counter-Reformation	Hanseatic League, trade in Northern and Western Europe
Bodin's absolutism (1576), Cromwell, Hobbes's *Leviathan* (1651)	Dutch Republic (1568), with democratic individualistic Protestantism (Aristotelian–Scholastic) (1618)
Louis XIV in France	William and Mary, the Glorious Revolution in England
Leviathan revisited: Rousseau (1772) Collective welfare principle: Bentham (1776)	Independent economy: Smith (1776)
French Revolution – Elements of liberalism: the common good defined in relation to the absolutist state	Diverse – Elements of liberalism: limiting domains of state power
Other "isms": socialism, communism, right-wing libertarianism as the strongest prevails, left-wing libertarianism as all are collective but local	But: Marx/Engels, the "withering away of the state" to occur
Democracy by representation	Limits of majority decision making: beyond Arrow (1951)
"We the people," direct democracy, referenda; UN Millennium and Sustainable Development Goals	Limited domains of power, specialized Independent economic institutions "Keynes": Central banks; IMF, GATT/WTO
Collective welfare scholars: Samuelson, Sen, Nussbaum, Sunstein, Stiglitz	Institutionalist social scientists/historians: North, Mokyr, Acemoglu, Ostrom

Note
GATT, General Agreement on Tariffs and Trade; IMF, International Monetary Fund; WTO, World Trade Organization.

governance are always being actively created and developing. But an atavistic resurrection of old ideologies may also occur. A return to 18th-century mercantilist nationalistic views is possible politically, as such views are still broadly present in many countries. This reversion to older approaches would restrict the institutions which are inductive to globalization and economic growth, closing borders and creating international tensions. This would lead to economic contraction and zero-sum international competition, neither of which is conducive to national climate policy or to international cooperation on climate policy.

Institutions, optimality and fissures after WWII

Post-WWII institutions and economic growth

The post-WWII explosion of global economic growth was characterized by unity in political philosophy regarding the economy, the Institutionalist version. This growth spurt in industrialized countries was guided by novel ideas about how national and international society should develop in the aftermath of the Great Depression and the atrocities of WWII. The business cycle was brought substantially under control based on insights into macro-level Keynesian mechanisms of supply and demand. The strict gold standard, which led to large and frequent depressions, was replaced by a system involving active control of the money supply to ensure price stabilization, combined with fiscal policy to regulate aggregate demand. Education was made available, largely free of charge, to the masses. Research, publicly funded for the most part, became the backbone of innovation for industrialization. Infrastructure was supplied publicly to allow for economic expansion. Natural monopolies in the areas of transport, energy, water supply and sanitation became public services, expanded for the common good. New near-monopolies as, for example, IBM in computers were split into smaller units. Where private firms could not deliver results, as in setting up new industries, public planning aided the process, most often related to required infrastructure and spatial planning. Public housing programs substantially resolve backward 19th-century housing conditions of the unschooled working classes. Emancipation of the working classes was also privately organized, especially by trade unions, which were supported by the state in this task. Universal suffrage, achieved around World War I (WWI), was becoming organized in grassroots organizations linked to political parties. At the international level, the common good was safeguarded by the IMF in terms of international monetary stability and growth options, by GATT/WTO in avoiding trade wars and opening up trade, by the UN with respect to security issues, and by the large military blocks of NATO and Warsaw Pact countries, which kept each other at bay. The international "hate-and-rob" actions that followed WWI, which Keynes identified in 1919 as predictive of the next world war (Keynes, 1920/2004), as well as the economic wars of the 1930s, including a breakdown of international trade after passage of the Smoot-Hawley Tariff Act in the United States (1930), were all successfully avoided following WWII. New institutions for the Western OECD countries stimulated unprecedented global economic growth, and a next wave of such growth starting in the 1990s moved China and other developing countries toward globalized industrialization.

At nearly the same time, however, the overall goal of welfare maximization – the old Bentham adage of the greatest happiness for the greatest numbers – became operational as a result of the ideological and scientific developments of the time, and availability of more detailed data and computing power. Now, large-scale institutional measures could be analyzed in detail, showing options for further improvement at that detailed level, similar to the way firms analyzed their operations in detail to reduce costs and increase the quality of their products. Building regulations, social welfare differentiations, energy standards for an ever-growing number of products, differentiated environmental regulations, and much more spread out a web of direct regulations. This process is ongoing, although incidentally reorganized toward simplicity in an administrative-technical manner: as Cass Sunstein advocated, simpler is the future of government (Sunstein, 2014).

Welfare maximization and domination of institutions for economic growth

Welfare maximization was the ideological goal of virtually all reasoning for these ever more detailed planning oriented developments, in addition to forming a part of economic theory. Mathematical developments in economics were accompanied by formalized welfare theory, developed around WWII, with Bergson and Samuelson as central figures (Samuelson, 1977). Bentham's goal became operationalized at several levels, with well-defined public tasks creating well-defined room for firms to grow and create private prosperity and thereby contribute to public prosperity. Operational welfare theory guided all public action, not from the perspective of organized private interests, but from a broader societal view, including non-market elements of welfare. The ultimate welfare criterion is the private utility of consumers, not of firms or their owners. The most operational welfare tool developed was cost-benefit analysis, which was applied to specific public actions ranging from new regulations to infrastructure development and made obligatory in the public domain. Cost-benefit analysis ostensibly reveals what is best for the people, taking into account market imperfections in complex models. Where market deficiencies existed, such as external negative effects not taken into account by firms and consumers, correction mechanisms were reasoned from welfare theory: the cost of a preventive measure should be lower than the damage caused. Environmental policy in the 1970s was contested in the details but enjoyed support along general lines across the full political spectrum, as all parties reasoned from the same welfare considerations. In the United States, for example, President Nixon – a mainstream Republican representing "the establishment" – set up the Environmental Protection Agency (EPA). And it worked. Economic growth in the OECD-countries took off for all in a manner unprecedented in history. Incomes rose, income inequality first diminished, intergenerational mobility expanded and slums and poverty receded.

Historians of different persuasions broadly agree that institutions are what drive societal development, most often actively established on the basis of general values and goals. Institutional analysis follows from general political philosophy. Around 1970, a tension arose within the broadly accepted democratic welfare approach. At its heart was a question of direction: whether to improve details ever further, implicitly move away from the institutionalist orientation that started the process, or accept diversity and mend the most obvious dysfunctionalities in the institutional framework. Among scholars, Samuelson, Sen, Nussbaum, Stiglitz, and Sunstein tend to support the ever more detailed collective welfare approach, whereas North, Mokyr, and Acemoglu represent the more generic institutionalist approach, but having the same welfare orientation. This tension was most evident in the high level of global economic growth generated by institutions following WWII, while at the same time political philosophy became ever more detailed regarding how to actively enlarge the common good. This evolution was most clearly visible around 1970 with the successful development of environmental policy, primarily through the use of detailed permits per installation. The economic welfare theory of the 1960s and 1970s could be used to rationally plan environmental policy, using cost-benefit analyses and public infrastructure such as roads, railways and airports. Recognition of its deficiencies in failing to account for distributional and ethical issues as well as personal development, led to improved welfare theory (Nussbaum & Sen, 1993). But these improvements also made welfare theory less operational. The recognition of widespread market failure, the subject of the Nobel Prize for Akerlof, Spence and Stiglitz, further detracted from its general operationality (Lofgren, Persson, & Weibull, 2002). Improving from fourth-best options to third and second best on the road to optimality does not have a defined path or well-defined outcome. Improvements on specific issues lead to ever more detail. Requiring building insulation in sunny, shielded surroundings is less useful than in colder windy surroundings, making requiring a differentiation in such regulations. Next, these regulations can be differentiated as to the housing class to which they refer: low-priced buildings for lower income groups need special attention in Planning & Control climate policy to alleviate energy poverty. Although aimed at concrete improvements, the welfare analysis has become partial and eclectic: the overall integrative framework is no longer operational. This makes it difficult to consider diffuse system effects, where changes in one rule require adaptations in other rules. The desired optimum is lost in a diversity of reasoning and options, with complexity making integration of different policy domains increasingly difficult. The basis of unified democratic welfare liberalism has become shaky. In this situation climate policy developed, with partial and detailed measures such as feed-in tariffs and emission standards making somewhat more generic options such as tradable emission permits ineffective.

Increased complexity disturbs climate policy

This divergence in approaches became more visible after approximately 1980. The opening of international institutions for economic growth to developing countries on the one hand, as developing countries including China moved internally toward open markets on the other, induced a second round of unprecedented global economic growth. This surge in growth lifted billions out of poverty in the last three decades. For the purposes of economic growth, establishment of the GATT/WTO followed the institutionalist approach. At the same time, the political philosophy based on democratic welfare theory shifted toward ever more detail in terms of goals and means and the detailed instrumentation needed to reach them. Exemplifying this are the 17 UN SDGs, which are specified in terms of 230 operational targets, each requiring its own instruments, to be aligned nationally and internationally. For the climate goal, this integrative alignment is described under the heading of mainstreaming climate issues in all policies (Kok & De Coninck, 2007; Rauland & Newman, 2015; UNDP, 2012).

The economic scores achieved under Institutionalism are impressive. In contrast, the climate scores associated with the current, mainly Planning & Control, orientation are meager, with CO_2 concentrations in the atmosphere continuing to rise relentlessly. The SDGs with respect to poverty and economic development tend to the concrete, using measures to alleviate poverty as in development aid. However, the massive reduction in poverty of the last quarter century has not been accomplished by aid programs, but by opening up Western international institutions to developing countries and by their internal institutional adaptations, creating markets in major domains of the economy. This institutional revolution is much criticized for its often harsh, purely economic orientation.

Proposed solutions take two very different approaches. One option is to differentiate rules of trade so as to incorporate other welfare aspects in them, as in differentiating rules relating to products with climate importance or underdeveloped, poor countries. (The discussions in the Doha Round of the WTO, effectively halted by the President Obama administration, reflected this approach.) The other option is improve *other* institutions, for example by focusing on education systems, reduced private monopolies, the tax structure and a differentiated power structure, safeguarded in a stable constitution. In terms of governance, the climate problem seems to require a more institutionalist approach as a viable alternative, perhaps more viable than pursuing more detailed and stringent regulations. But other issues loom in the background, requiring a broader review of governance and political philosophy.

Increases in inequality disturb climate policy

Climate change is not the only issue requiring institutional change. The economies of developed Western countries not only are growing more slowly than those of the newly developing countries, but also have experienced an extremely unequal growth in primary incomes during roughly the past three or four decades. The lower 50% in primary income distribution have seen hardly any rise in income while the richest 50% have received nearly all of the income and wealth increase. A substantial share accrues to the richest 10%, with the greatest share going to the richest 1%. Wealth of the top 0.1% percent has exploded. Globally, the picture is very different, as the middle income levels, dominated by developing countries, have exploded (see the elephant curve in Figure 2.1, earlier). The stagnating income at the neck of the elephant covers the ninth decile globally. However that 10% globally constitutes the 50% lowest income groups in developed countries. This extreme growth in income inequality there is also increasing within developing countries, with an even more extreme rise in inequality of wealth. The rise in national inequality is not only an economic issue but a political issue as well, likely contributing to new forms of populism that are not helpful for enacting deeply effective climate policy. Excluding a large part of the population from the benefits of economic growth certainly does not fit with overall welfare improvement, as one dollar going to a rich person creates less welfare increase than one going to a poor person, according to welfare theory.

Several mechanisms would seem to contribute to the cause of these unequal results.

- In developed countries, one mechanism is the competition coming from newly developing countries. The new middle classes there outcompete less-educated persons in developed countries. But, for the most part, the mechanisms seem similar at a global level. Technological development tends toward decreasing marginal costs. Software that is developed can be provided at hardly any additional costs. Information exchange is more efficient using one system that connects all. One or a few firms will remain, reaping monopolistic benefits. Internet and cheap transport make markets and new technologies global, with global near monopolies emerging.
- The so-called Second Machine Age (Brynjolfsson & McAfee, 2014) reinforces this direction by absorbing more routine intellectual tasks into IT programs developed by highly educated specialists, and moving toward AI.
- Patent law protection reinforces a tendency toward oligopoly and temporary monopoly, with the number of patents rising astronomically in even more domains. In densely covered patent areas, new and smaller firms simply cannot operate, while larger firms spend a substantial part of their development budget on patent creation and patent protection, reducing freely available general knowledge in the process. Universities join in this development by privatizing research results using patents and participating in spinoffs. The justification for patent protection comes from the welfare school as patents are assumed to support innovation. As previously noted, James Watt helped develop the modern patent system to protect his innovative steam engine. A specific patent may indeed pay the patent owner substantially and possibly create an incentive for technology development. But at the same time, the thicket of several million patents, utility models and protected industrial designs[3] reduces overall innovation. The usable knowledge which formerly used to be available freely has become proprietary to a high degree in the past few decades, with firms competing in patents more than in design (Boldrin & Levine, 2008; Stiglitz, 2014). The use of patent trolls is on the rise, to cash in on the innovations of others.

The second and third mechanisms, both apart and together, create a tendency toward monopoly whereby the winner takes all. Next, billionaires can use their political power to protect their substantial interests, including better protection of the patents they own.

Such major developments must be reckoned with in designing climate policy instrumentation, be it from a Planning & Control or an Institutionalist perspective. They comprise the socio-political context in which climate policy should effectively come about. The solutions sought will, however, differ deeply, depending on the branch of political philosophy from which they spring. The integrative approach in Planning & Control requires climate policy to tackle other major problems as part of climate policy, in ever better detail. In the Institutionalist perspective, climate policy instrumentation is set up to be as independent as possible from other institutional domains, leaving the solution of other problems to be handled as independently as possible, as well.

Unity and later fissures in liberal political philosophy

The development of political philosophy in Western democracies since the 17th century has centered on several forms of liberalism, in the broader sense of that term. New *isms* – socialism, communism, anarchism, and libertarianism – presented themselves in the 19th century, arising from the same liberal root. The extremes have remained at the fringe or have fallen by the wayside. The Spanish Civil War dealt a defeat to anarchism in the 1930s. Socialism was transmuted into social democracy, represented by the social democratic liberals in Europe and, similarly, the liberal wing of the Democratic Party in the United States. National Socialism was defeated in WWII. Communism succumbed to oligarchic capitalism in the USSR and was transformed into a semi-liberal system of state capitalism in China. Daniel Bell declared the end of ideology in 1962 (Bell, 1962). The end of history was declared by Frances Fukuyama 30 years later (Fukuyama, 1992). Global liberal institutions, dominated by the United States since WWII, would bring freedom, order and growth throughout the second half of the 20th century, guided by welfare theory and democracy. That was the phase of unity, but fissures later appeared when welfare theory lost its power of persuasion, centrality and undisputed operationality.

Economic welfare reasoning remained, but lost much of its persuasiveness and centrality – this was true in the newly emerging climate policy domain as well. The costs and benefits of climate change and climate policy were quantified in the old welfare fashion in the Stern report (Stern, 2008). But quantifications of damages caused by CO_2 emissions are widely contested; see Tol's acrimonious survey (Tol, 2008). Results diverge more than one order of magnitude, all based on reasonable methods and reasonable assumptions. Worse, they leave out the major effects of catastrophic climate change, which is not amenable to economic quantification (Weitzman, 2011). Increasingly policy is less based on welfare reasoning and more influenced by disconnected emotional issues such as the fear of disaster, historical justice, international equity, national justice, and social independence in restricted communities – all relevant but hardly reducible to monetary values, and certainly not open to trade-offs for optimization. Welfare theory and its liberal ideology are no longer linked to clear solutions of generally defined problems.

This change is reflected in the different direction later developments have taken, starting at the end of the 1970s with the new conservative neo-liberal views expressed in the Reagan–Thatcher reforms after Western economic growth had reached its height. It can also be seen in recent electoral losses of mainstream liberal parties, and gains of diversely defined new parties and policy views. Formally, however, welfare-oriented, economic optimality reasoning remains intact. In the United States, for example, public policies are still required by law to undergo cost-benefit analysis. President Reagan's E.O. 12291, mentioned earlier, is an example, with later improvements to reflect increased complexity. E.O. 12688, signed in 1993, stated that the U.S. federal government can make regulations as "… are made necessary by compelling public need, such as material failures of private markets to protect or improve the health and safety of the public, the environment, or the well-being of the American people." Twenty years later a call was made for an executive order requiring better-integrated analysis, through "deeper and wider cost-benefit analysis" (Hahn & Sunstein, 2002). However, by then that option had more or less ceased to exist, with Sunstein shifting to advocate simplification for the future of government (Sunstein, 2014).

As a global problem climate change requires some international coordination, preferably globally, as exemplified in the UN SDGs, of which a stable climate is a part. However, as Samuel Huntington argues, the level of global unity that was dominant ideologically after WWII is evaporating, in a move away from global liberalism to culturally defined identities with nationalist and religious overtones (Huntington, 1996). His *Clash of Civilizations* was favored reading by both Osama Bin Laden and George W. Bush and certainly would not seem to provide fertile soil for defining the global common good, let alone creative global and national climate policy instrumentation.

Applied welfare theory was transformed into ever more complex specification of goals and targets, but at an operational level covering only limited and partial reasoning. Public policy lost its uniform direction and appeal. The post-WWII period of innovation, creativity and inspiration broke down in Western countries. Striving for the overlapping goals of democratically based increases in individual and collective welfare could not be maintained in practice. This breakdown created an opening for partisan interests to manipulate the political and administrative system, while continuing to use the terminology of old that focused on the best for all, and having regard for all that is important. The social democratic liberal trinity of *liberté*, *égalité* and *fraternité* is crumbling in many Western countries, evolving into liberty for some, highly unequal distribution of income wealth and power, and the loss of social cohesion. This change reduces options for effectively addressing any long-term and complex policy issue, but certainly the "wicked climate problem" (Levin, Cashore, Bernstein, & Auld, 2012), which has extremely complex, globally interrelated causes. The climate problem is fully global in its effects, and cannot be resolved by a simple, single measure. Social and political instability and incoherent political philosophy are not the friends of effective climate policy.

Summary of inequality and crumbling welfare philosophical unity

- Throughout the latter half the 20th century, welfare maximization was supported by applied welfare theory, but welfare theory collapses under the overload of goals, targets and the complexities in application, leading to endless numbers of partial regulations. Freedom from want was realized, but broader freedom was restricted by increasingly dense and complex top-down regulations. Climate policy, and other policies as well, drifted from agreed welfare maximization to multiple, single-issue actions, vulnerable to political interventions by partial interests.
- Continuing this trend, democratic optimality lost its operational basis in applied welfare-theory. The reduced operationality of integrated welfare reasoning is indicated by now specifying independent welfare domains (Jones & Klenow, 2016) and similarly can be derived from the inhomogeneous nature of the UN Sustainable Development Goals and their large number of non-integrated operational targets.
- The goal of collective progress is succumbing to partial interests and issue politics, while discussion of issues tend to be dominated by the extremely rich and powerful in an often antagonistic way.
- Increasing equality in the aftermath of WWII has now risen to levels not seen for a century. The pattern may be well-known historically, but the mechanism and speed are new. Both the super-rich and the class of newly poor and disarrayed poor may not have much interest in climate policy, or in any broader policy approach.
- A clear basis for climate policy instrumentation in these dynamic and diverging developments in political philosophy, political practice and broad socioeconomic trends is no longer present. Therefore, there is no clear starting point for effective climate policy instrumentation. Moving beyond the current idea of democratic welfare, governance seems unavoidable
- The two forms of reasoning that have long been present need a new definition: genuinely effective detailed technology and market regulations in Planning & Control, and fundamentally revamped institutions in Institutionalism.

Climate policy within liberal democracy

We the people *as a constitutional solution?*

Would we have to leave the domain of liberal representative democracy? But what viable alternative might replace it that would leave most of us alive and happy? Democracy is widely acclaimed as the constitutional core of Western society, without any serious alternative. There are variants and details, but not an alternative. The democratic welfare mode of governance still dominates public policy reasoning toward the common good, based on democratic centralization and guided by the idea of integrated welfare theory. Measures for the common good, as democratically conceived, must be implemented in the most efficient way, as well as procedurally. But what constitutes the common good and what the best means are for realizing it have become unclear. Increasing income inequality and reduced social welfare support show the lack of power of democratically chosen majority governments. This outcome certainly is not what *we the people* want. The democratic deficit is widely felt, described as diminishing the legitimacy of government. This development has been measured in detail for the countries of the EU (Blanco-González, Prado-Román, & Díez-Martín, 2017). One constitutional solution is to shift from representative government toward direct democracy using internet-based consultations and an increasing role for referenda, with the latter shifting from consultative to binding – if not legally then in practice, as with Brexit. *We the people* may thus be interpreted in a very direct way. Another option for constitutional change is to more directly reflect broad opinion in society by *not* having elections and referenda and instead selecting representatives by lot, somewhat in line with Aristotle's approach in the *Constitution of Athens* (Reybrouck, 2016).

It seems, however, that it is not the serious flaws in democratic decision-making procedures that cause lack of legitimacy, but the deeper incapacity to define and implement the ideal of the common good. The solution cannot be direct democracy, which would reduce complex political issues ever further in an arbitrary way to single yes-no decisions. This would open them up to more manipulation than is presently the case through current manipulation of elections. Direct democracy will fail in complex situations where there are no single choices to further the common good. Climate policy is one clear example of such a complex issue.

Broader coalitions, with competing political philosophies?

Cracks and fissures are now visible within the democratic, welfare-based, liberal mainstream of industrialized countries. In a dialectic way this demise could lead to the emergence of more effective Planning & Control or to a more Institutionalist development. But is a new approach also possible politically, linking different strands of political philosophy? Candidates from other "isms" include left-wing and right-wing libertarianism, communism, and, less clearly defined, conservatism. Several versions of anarchist left-wing libertarianism exist. Nozick may be seen as the ideologue of the left-wing version of libertarianism, with small movements of a more active political nature such as "degrowth" (D'Alisa et al., 2014; Kallis, 2011; Nozick, 1974). Anarchist left-wing libertarianism subscribes to the collective ownership of production, but on a local scale only. Climate considerations may be included in local decision making. However, the local groups can hardly be considered partners in the climate coalition of a large country. The right-wing libertarian wing of Ludwig von Mises' Austrian School reduces the tasks of the state to guaranteeing security, the most extreme version being Rothbard's anarcho-capitalist philosophy in which even the police function is privatized. The anarcho-capitalist version has a stronghold in the United States, mainly linked to the Republican Party (Mack, 2015; Rothbard, 2009; Block & Rockwell Jr, 2007; Hoppe, 1999). Climate coalitions with the liberal political center do not seem feasible as for them it is not a function of the state to deal with climate change.

Opposed to anarchist libertarianism is centralized communism, striving for the long term good of all, now mainly at a national level In China, the main country in which a communist party is still in power, substantial state-capitalism and direct control of private enterprise are part of "Xi-ism." The link between big business and government is quite direct, however, easily linking public policy to these private interests, similar to developments in the United States. In China both tradable emission permits and emission taxes are being employed or considered for climate policy, alongside direct regulation. Coalitions are possible in multi-party systems as communism has the same foundation, although interpreted quite differently, in striving for the common good, long term. In some industrialized countries a small communist party in favor of climate action in principle exists, but such parties are hardly relevant politically. This leaves the basket of the 19th-century "isms" quite empty. Anarchists, libertarians and communists either do not want to or cannot play an active role in developing climate policy or its instrumentation in the liberal domain. However, an agreement at an international or global level is possible with a communist country such as China, assuming long-term climate policy will develop there for the good of the nation.

Conservatism may have several sources, including linking to a romantic past or to stability, or both. Bismarck-type conservatism views stability for the nation as a prime goal, requiring economic development for reasons of national legitimacy and international economic competitiveness and power, similar to Xi-ism. There is no direct link to democratic welfare, however, placing this philosophy only just inside the broadly defined liberal domain. The modernizing version may be reckoned as part of broadly defined liberalism; the romantic type, not. The political philosophies of libertarianism, communism and romantic conservatism diverge from the core of welfare-oriented liberalism. Nonetheless, there is an opening to institutionalist climate policy for Bismarck-type conservatism.

Coalitions within liberalism seems to be the prime option for a broadly supported climate policy of any kind. One option is to strengthen the current, mainly Planning & Control, majority. In all Western countries this requires coalitions, and also participation of the United States. Despite a Democratic president who advocated for action on climate change for 8 years, the United States did not reduce its CO_2 emissions and expanded its fossil fuels production. Subsequently, the U.S. Republican government reversed course, undercutting prior commitments to expand renewable energy and curtail CO_2 emissions. Stricter regulations and higher subsidies may be possible to some extent, but only if a broad spectrum of political groups is involved. In the United States, this will require the support of at least a substantial segment of the Republican Party, and overcoming resistance to substantially deeper regulation in the Democratic Party. It is not just a simpler regulation, as proposed by Sunstein, but substantially more and more stringent regulation. Bismarckian conservatives will not join in more stringent planning, nor will those whose political philosophy is closer to libertarianism. In seeking to implement more detailed regulations, a link with Christian Democracy may be possible through its focus on local social stability and inclusiveness as a requirement of Christian brotherhood (Etzioni, 1996). Deeper regulation does not just represent a bit more of current policy instrumentation, however, but a fundamental shift toward effectiveness, phasing out major industries, and bringing in new ones in an encompassing planned way, much beyond good intentions. The problem of climate change is clear, it is broadly accepted in almost all political directions, also in the US.

A winning coalition in Institutionalism must overcome two hurdles. At the left and center it must overcome the centralizing and integrating tendency of doing practical things, now. The shift to Institutionalism must be seen as a logical consequence of desired climate effectiveness, avoiding the practical action view that easily leads to the current clash of mutually adverse instruments. More stable is a shift resulting from a reorientation of social democracy and broader left and center liberalism in the Aristotelian direction. Attaining a majority will require active participation of also the conservative segments of liberalism, including elements of conservative neo-liberalism. Bismarck-type conservatism is a possibility in the Republican Party and some Christian denominations in the United States, and the somewhat related Christian Democrats in Europe. Their ideas regarding a stable community derive from the notion of brotherhood expressed in the French Revolution – a notion that has receded in modern capitalism, where the focus is on personal and economic freedom and on equality of opportunity. Bismarckian conservatives see climate change as a danger to stability that must be acted upon; they abhor detailed regulations and for such administrative-technical reasons align more easily with Institutionalism. Conservatives within the U.S. Republican Party have proposed active climate policy that uses an encompassing emission tax; one example is the CO_2 emission tax outlined by former Secretary of State James Baker in a pamphlet entitled "The conservative case for a carbon dividend" (Baker et al., 2017). This branch of conservatism has a nationalist overtone, and therefore does not link easily to international action. The Bismarck-type and Christian Democratic versions of conservatism may present options for forming coalitions that would differ from the two branches of welfare liberal climate governance distinguished here, with the Bismarck type linked to Institutionalism and the Christian Democratic type more likely to support Planning & Control, but not necessarily so.

These diverse directions in political philosophy are summarized in Figure 2.2.

Climate policy at the philosophical crossroads within liberalism

The old distinction in political philosophy – governance through central goals and efficient implementation versus governance through decentralized autonomy

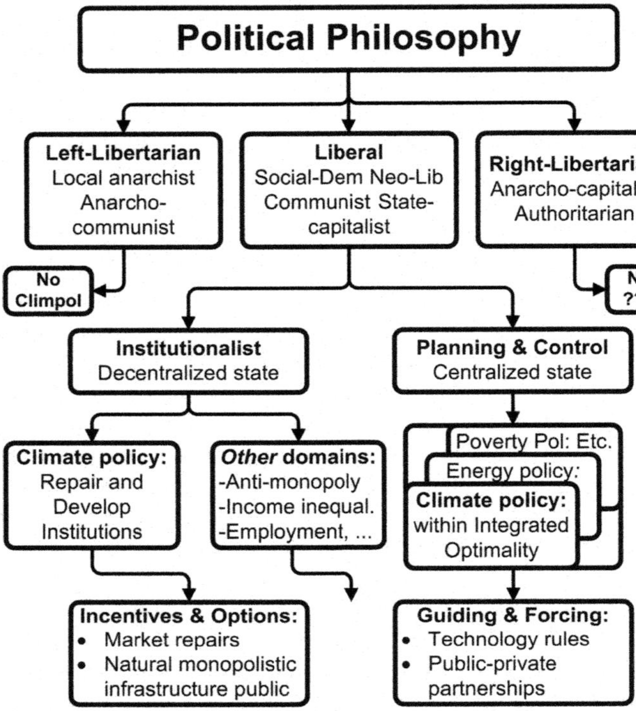

Figure 2.2 From political philosophy to governance.

within safeguarding institutions – has not lost its relevance for today's world. The centralist approach, represented here as Planning & Control, is losing ground as democratic optimality has lost its operational democratic welfare basis due to the normative and empirical complexity, resulting from the enriched coverage of welfare theory in an ever more complex global economy. Revitalization of this line of governance and instrumentation is required to achieve effective climate policy. This is the first direction at the philosophical crossroads for climate policy instrumentation, detailed in Chapter 4. It expands outward using much of current instrumentation, but then becomes more coherent, more centralized, more forceful, and also smarter and simpler. The decentralized autonomy approach, represented here as Institutionalism, refers back to the rules and mechanisms that have created modern industrialized society. It consists of a shift to new policy, based on revised institutions that guide decentralized decision making toward deep emission reduction by creating incentives and options. This is the second direction at the philosophical crossroads for climate policy instrumentation. Not making that crossroads choice explicitly leaves policy instrumentation without guidance, and can easily lead to the ineffective clash of partial instruments such as currently exists.

Notes

1. A shorter and easier accessible survey can be found at https://en.wikipedia.org/wiki/Republic_(Plato).
2. Aristotle's text can be downloaded from http://classics.mit.edu//Aristotle/athenian_const.html.
3. These statistics are available at http://www.wipo.int/ipstats/en/charts/ipfactsandfigures2016.html.

References

Acemoglu, D., & Robinson, J. A. (2012). *Why nations fail: the origins of power, prosperity and poverty*. New York: Crown Publishers.

Aristotle. (330–322 BC). Constituton of Athens. (1920), F.G. Kenyon transl. In J. Barnes (Ed.), *The complete works of Aristotle. The revised Oxford translation* (vol. 2, pp. 2341–2383). Princeton, NJ: Princeton University Press.

Aristotle. (1950). *Aristotle's Constitution of Athens and related texts. Translated with Introduction and Notes by Kurt von Fritz and Ernst Kapp* (K. von Fritz & E. Kapp Eds.). New York: Hafner Press, McMillan Publishing Co.

Arrow, K. J. (2012). *Social choice and individual values*. New Haven, CT: Yale University Press. (Originally published 1951.)

Baker III, J. A., Feldstein, M., Halstead, T., Mankiw, N. G., Paulson Jr, H. M., Shultz, G. P., … Walton, R. (2017). *The conservative case for carbon dividends*. Retrieved from https://www.clcouncil.org/.

Bell, D. (1962). *The end of ideology: on the exhaustion of political ideas in the fifties*. Cambridge, MA: Harvard University Press.

Bentham, J. (1776). *A fragment on government: being an examination of what is delivered, on the subject of government in general.* London: Payne, Elmsly and Brooke.

Blanco-González, A., Prado-Román, C., & Díez-Martín, F. (2017). Building a European legitimacy index. *American Behavioral Scientist, 61*(5), 509–525. doi: 10.1177/0002764217693282

Bloch, M. (1961). *Feudal society* (2 volumes, transl. L.A. Manyon). Chicago: University of Chicago Press. (Originally published 1939.)

Block, W., & Rockwell Jr, L. H. (2007). *Man, economy, and liberty: essays in honor of Murray N. Rothbard.* Auburn, AL: Ludwig von Mises Institute.

Bobonich, Chris and Meadows, Katherine, "Plato on utopia", *The Stanford Encyclopedia of Philosophy* (Spring 2018 Edition), Edward N. Zalta (ed.), URL = <https://plato.stanford.edu/archives/spr2018/entries/plato-utopia/>.

Bodin, J. (1992). *Bodin: on sovereignty*. Cambridge, England: Cambridge University Press. (Originally published 1576.)

Boldrin, M., & Levine, D. K. (2008). *Against intellectual monopoly*. Cambridge, England: Cambridge University Press.

Brown, E. (2011). Plato's ethics and politics in The Republic. In E. N. Zalta (Ed.), *The Stanford encyclopedia of philosophy*. Retrieved from http://plato.stanford.edu/archives/win2011/entries/plato-ethics-politics/.

Brynjolfsson, E., & McAfee, A. (2014). *The second machine age: work, progress, and prosperity in a time of brilliant technologies*. New York, NY: Norton.

D'Alisa, G., Demaria, F., & Kallis, G. (2014). *Degrowth: a vocabulary for a new era*: New York: Routledge.

Etzioni, A. (1996). The responsive community: a communitarian perspective. *American Sociological Review*, *61*(1), 1–11. doi: 10.2307/2096403
Fukuyama, Y. F. (1992). *The end of history and the last man*. New York: Free Press.
Giddens, A. (1998). *The third way: the renewal of social democracy*. Cambridge, MA: Polity Press.
Gimpel, J. (1977). *The medieval machine: the industrial revolution of the Middle Ages*. New York: Penguin Books (French original La Révolution industrielle du Moyen Age, 1975).
Hahn, R. W., & Sunstein, C. R. (2002). A new executive order for improving federal regulation? Deeper and wider cost-benefit analysis. *University of Pennsylvania Law Review*, *150*(5), 1489–1552.
Hilton, R. H. (1995). *English and French towns in feudal society: a comparative study*. Cambridge, England: Cambridge University Press.
Hobbes, T. (1968). *Leviathan*. C. B. Macpherson (Ed.). Harmondsworth, England: Penguin Books. (Originally published 1651.)
Hoppe, H.-H. (1999). Murray N. Rothbard: economics, science, and liberty. In R. G. Holcombe (Ed.), *Great Austrian economists*. (pp. 223–241). Auburn, AL: Ludwig von Mises Press.
Huntington, S. P. (1996). *The clash of civilizations and the remaking of world order*. New York: Simon and Schuster.
Jones, C. I., & Klenow, P. J. (2016). Beyond GDP? Welfare across countries and time. *American Economic Review*, *106*(9), 2426–2457.
Kallis, G. (2011). In defence of degrowth. *Ecological Economics*, *70*(5), 873–880.
Keynes, J. M. (2004). *The economic consequences of the peace*. Mineola, NY: Dover. (Originally published 1920, London, MacMillan.)
Kok, M., & De Coninck, H. (2007). Widening the scope of policies to address climate change: directions for mainstreaming. *Environmental Science & Policy*, *10*(7), 587–599.
Levin, K., Cashore, B., Bernstein, S., & Auld, G. (2012). Overcoming the tragedy of super wicked problems: constraining our future selves to ameliorate global climate change. *Policy Sciences*, *45*(2), 123–152.
Lofgren, K. G., Persson, T., & Weibull, J. W. (2002). Markets with asymmetric information: the contributions of George Akerlof, Michael Spence and Joseph Stiglitz. *Scandinavian Journal of Economics*, *104*(2), 195–211. doi: 10.1111/1467-9442.00280
Mack, E. (2015). Robert Nozick's political philosophy. In E. N. Zalta (Ed.), *The Stanford encyclopedia of philosophy*. Retrieved from http://plato.stanford.edu/archives/sum2015/entries/nozick-political.
Maskin, E., & Sen, A. (2014). *The Arrow impossibility theorem*. New York: Columbia University Press.
Milanovic, B. (2016). *Global inequality: a new approach for the age of globalization*. Cambridge, MA: Harvard University Press.
Nozick, R. (1974). *Anarchy, state, and utopia*. New York: Basic Books.
Nussbaum, M., & Sen, A. (1993). *The quality of life*. Oxford, England: Oxford University Press.
Ostrom, E. (2010). Beyond markets and states: polycentric governance of complex economic systems. *American Economic Review*, *100*(3), 641–672. doi: 10.1257/aer.100.3.641
Piketty, T. (2014). *Capital in the twenty-first century*. Cambridge, MA: The Belknap Press of Harvard University.
Popper, K. S. (2012). *The open society and its enemies*. *Volume I: the spell of Plato*. London: Routledge. (Originally published 1945, reprinted 1966.)
Rauland, V., & Newman, P. (2015). *Decarbonising cities: mainstreaming low carbon urban development*. Cham, Switzerland: Springer International.

Rawls, J. (2009). *A theory of justice*. Cambridge, MA: Harvard University Press. (Originally published 1971.)

Reybrouck, D. v. (2016). *Against elections. The case for democracy*. London: The Bodley Head/Penguin Random House. (Dutch original 2013.)

Rothbard, M. N. (2009). *Man, economy, and state*. Auburn, AL: Ludwig von Mises Institute. (Originally published 1971.)

Rousseau, J.-J. (2008). *Discourse on political economy and the social contract*. Oxford, England: Oxford University Press. (Originally published 1772.)

Samuelson, P. A. (1977). Reaffirming the existence of "reasonable" Bergson-Samuelson social welfare functions. *Economica*, *44*(173), 81–88. doi: 10.2307/2553553

Scheidel, W. (2017). *The great leveler: violence and the history of inequality from the Stone Age to the twenty-first century*. Princeton, NJ: Princeton University Press.

Schumpeter, J. A. (1949). Vilfredo Pareto (1848–1923). *Quarterly Journal of Economics*, 147–173.

Sen, A. K. (1970). *Collective choice and social welfare*. San Francisco: Holden-Day.

Smil, V. (2009). *Global catastrophes and trends: the next fifty years*. Cambridge, MA: MIT Press.

Smith, A. (1979). *An inquiry into the nature and causes of the wealth of nations, Vol I & II*. Oxford, England: Oxford University Press. (Originally published 1776.)

Smith, M. (1904). Problems of Roman legal history. *Columbia Law Review*, *4*(8), 523–540. doi: 10.2307/1109104

Stavins, R. N., Wagner, A. F., & Wagner, G. (2003). Interpreting sustainability in economic terms: dynamic efficiency plus intergenerational equity. *Economics Letters*, *79*(3), 339–343. doi: 10.1016/S0165-1765(03)00036-3

Stern, N. (2008). The economics of climate change. *American Economic Review*, 98(2), 1–37. doi: 10.1257/aer.98.2.1

Stiglitz, J. E. (2014). *Intellectual property rights, the pool of knowledge, and innovation*. (Working Paper No. 20014). Cambridge, MA: U.S. National Bureau of Economic Research.

Stiglitz, J. E., Sen, A. K., & Fitoussi, J.-P. (2009). *The measurement of economic performance and social progress revisited: reflections and overview*. Paris: OFCE. HAL. Id: hal-01069384.

Sunstein, C. R. (2014). *Simpler: the future of government*. New york: Simon and Schuster.

Tol, R. S. (2008). The social cost of carbon: trends, outliers and catastrophes. *Economics*, *2*(2008-25), 1–22.

UBS & PWC. (2017). *New value creators gain momentum. Billionaires insights 2017*. Retrieved from: https://www.pwc.com/gx/en/financial-services/Billionaires%20insights/billionaires-insights-2017.pdf

UNDP. (2012). *Mainstreaming climate change in national development processes and UN country programming: a guide to assist UN country teams in integrating climate change risks and opportunities*. New York: United Nations Development Programme.

Van Bavel, B. (2016). *The invisible hand? How market economies have emerged and declined Since AD 500*. Oxford, England: Oxford University Press.

Warde, P. (2007). *Energy consumption in England and Wales, 1560–2004*. Naples: Consiglio Nazionale delle Ricerche (CNR). Available at: https://histecon.fas.harvard.edu/energyhistory/data/Warde_Energy%20Consumption%20England.pdf

Weitzman, M. L. (2011). Fat-tailed uncertainty in the economics of catastrophic climate change. *Review of Environmental Economics and Policy*, *5*(2), 275–292. doi: 10.1093/reep/rer006

3 A guide for coherent design of climate policy instrumentation

Strategies for coherent instrumentation design

Design of instrumentation involves many disparate considerations. Individual instruments have to perform sufficiently in an administrative-technical sense; they must dovetail with other climate policy instruments, align with other policies and their instrumentation, enable international cooperation, should not be more costly than necessary, should not be vulnerable to interest-based interventions, should induce innovation of low-emissions technologies, should not merely shift emissions abroad, and as a mix of instruments the full instrumentation design should be capable of reducing emissions quickly and deeply enough to counter major negative effects. More specific considerations, such as regional employment, involving appropriate levels of government and compatibility with overall industrial development, also have to be made. Coherent instrumentation deals with all such relevant considerations, or places them outside of the climate policy domain as other independent subjects for policy. This increases the complexity of the design process. In addition, it is not clear how to define policy instruments in general, and environmental policy instruments more specifically. Climate policy instruments: what are they? An overall strategy is needed to navigate this bewildering multitude of divergent considerations. The two core directions in political philosophy can guide answers, giving direction to the analysis and the solutions. They both must be able to deliver the required deep emission reduction. That is the effectiveness requirement assumed here.

Enticingly simple solutions would seem to be available, at least in a technocratic sense. Closing all coal mines and all oil and gas fields by just 3% points per year would do the climate policy job. If started in 2020, reduction by 90% could be reached by 2050: 30 x 3% = 90%, and 3.3 years later no emissions would be left. Can it be that simple? Since they are the most polluting, why not first close coal-fired power stations by 6% per year, and reduce oil and gas next, also at 6% per year, and achieve roughly the same result by 2050? In terms of technical effectiveness, these options are both fine. But they are hardly fine in political and economic reality. Which mines and wells should be closed, and when? What should be done about imports? What about regions or countries dependent on oil, or coal, for income and employment? What if an ailing coal industry in a poor region asks for support? No political philosophy exists to steer society in this way, no mode of governance exists for this task. Politically speaking, very few national governments would be able to carry out this task, only those with little coal, oil and natural gas reserves. And no administrative systems are now available to deal with this task. There is also no central global government able to enforce such a scheme. Closure must be specified per coal-fired power station, but which one would go first, and then second? The economic consequences would be very difficult to oversee, but they would certainly lead to high costs, as equal speed for all fossil fuel reduction does not correspond to differences in cost and expected technical life time, with the current availability of other options for energy, nor for the energy reduction they will produce in time. Economically, it may be nearly impossible to follow this line, as massive investments would be required to expand materials production for renewable and low-emission energy-use systems, all requiring energy themselves for their investment, which would have to remain substantially fossil based for some time to come. The policy would be at variance with other equally desirable goals, such as economic growth and low unemployment. And, in order to start this way, political support would be required in almost all countries of the world, including fossil fuel–producing countries. Each issue arising along the way could block the road to the low emission target, even if the target is substantially agreed upon. Realistic design considers what is feasible. Coherent design involves taking account of what must remain fixed and what is malleable, all the while keeping the goal of climate effectiveness in mind.

Design discipline: layers of considerations

Several layers of considerations can be discerned in the design process of instrumentation fit for the climate stabilization goal as stated in the UNFCC Paris Agreement of 2015. The first layer, here referred to as Layer 1, involves the general governance setup of society: what it is, how it more or less autonomously develops, and how such development can be steered in a different direction. This layer of analysis includes the constitution of government and primary institutions, including the legal framework and systems of education, research and health care. This subject was treated in Chapter 1, and focused on two main lines in governance development: Planning & Control and Institutionalism. These were placed in a broader political theory background in Chapter 2.

Next, in Layer 2, is the socioeconomic system, with energy use, dynamics generating autonomous development, and options to steer away from less-desirable to more-desirable development. Prospective modeling focused on low-carbon developments was also surveyed in Chapter 1. Current broader issues concern continuing economic growth, growing income inequality, tendencies toward increasing monopolization, globalization of information and production, and, of course, ongoing, rapidly rising, global warming gas concentrations, CO_2 in particular. These more qualitative developments should also guide the details in instrumentation design.

Layer 3 is more directly focused on the climate issue, with emitting activities as the next objects of policy instrumentation in the economy. This layer again involves an analysis of autonomous development, with likely directions, and how this development may be influenced in economic-technical terms. Much of the technical-economic, scenario modeling by Intergovernmental Panel on Climate Change (IPCC) and International Energy Agency (IEA) takes place at this level, against a background of Layer 2 considerations. Most of these scenarios assume the effectiveness of some degree of instrumentation, often based on idealized emission pricing in idealized markets, abstracting from real-life implementation problems.

Layer 4 concerns the building blocks for instrumentation design. What instruments are possible, and how should they be specified? Ultimately, detailing to the level of administrative implementation is required. For example, the financial instruments of taxation and subsidy are often referred to as general instruments, including tax differentiations and levies, duties, excises and tariffs. Emission pricing is a subgroup within the group of taxation instruments, designed either as a specified tax, or through a cap on emissions by auctioning a limited supply of emission permits and making them tradable. Price levels, and public proceeds, then follow. Within the taxation subgroup, a tax may be encompassing and be equal for all emissions, or it may differentiate among domains, such as between transport emissions and electricity-generating emissions. And it may cover a proxy variable only, as in taxing energy or high-emitting cars instead of emissions. Such differentiations must be clarified in terms of how instruments are to function administratively, and also to avoid overlap and gaps.

The next layer of considerations, Layer 5, is the core subject of the analysis at hand: the composition of a coherent set of climate policy instruments. Coherency means instruments that are mutually supportive, cover all emissions, have little overlap, and in principle together can achieve the goal of limiting global warming to 2 degrees, or preferably 1.5 degrees.

The final layer of analysis, Layer 6, is reflective. All instrumentation choices are based on assumptions at each layer of reasoning and all involve substantial uncertainties. An additional question at Level 6 is: are the instrumentation designs resilient? Is the sum total of uncertainties to be dealt with low enough to remain on the safe enough side for climate stability? Can the lock-ins to existing technologies be overcome and can new technologies be implemented with presumed effectiveness? Is the instrumentation politically and administratively feasible and robust? This reflective analysis is the subject of the Chapter 6, following more detailed development of the two main policy directions for instrumentation in Chapters 4 and 5.

Table 3.1 provides a concise overview of design theory as related to the six layers of analysis.

Like any design process, whether it be for a car, a house or a new city, the design process for climate policy instrumentation is an abstract one, however practical it must ultimately be. It is easier to imagine a product design than an instrumentation design, and therefore product design can provide some inspiration and guidance here. In designing a future car, one does not start with the "instruments"

Table 3.1 Layers of analysis in designing climate policy instrumentation

1. General governance	Political philosophical directions, history and development
2. Economy and society	General autonomous developments, and active corrections
3. Emission reduction options	Long-term technical and economic options and scenarios
4. Climate instruments, single	Domain, types and detailing
5. Coherent instrument sets	Overall effective and efficient, governance dependent
6. Reflections on imperfections	Incommensurables, uncertainties and risks of choices, resilience

that make up a car — a steering wheel, a window or a door. Similarly, when designing a building, one does not start with the stones, doors or windows. It would also not be wise to start with the product structure, such as how to combine the motor, wheels, body and steering wheel. A level above the structure is the functionality of the whole, specified in terms of desired functionality defined in the broader context of the wider functioning of the product. The parts that together can deliver that functionality must be developed and selected in coherent detail for the more encompassing functionality task. A self-driving car may not need a steering wheel. Functionality of the car corresponds to the deep emission reduction here. But there is no single best new car design, and no single way for emission reduction. How is the functionality to be defined? How does this partial climate functionality relate to broader functionalities in society, which are also developing and of which the car design is part? Henry Ford's strategies were connected to an industrializing society with a well-paid workforce able to buy his cars, leading to a very specific design: the Model T. Current strategies at Ford Motor Company relate to the future of transportation, to a world where artificial intelligence will take over many tasks, including that of many a driver, and where the quality of urban life is at variance with private car systems with substantial parking requirements and emissions. A different car design will emerge, to be part of the broader transport system it is co-producing. Of course it is possible that this more general development may not come to pass in the way expected. Or the design process may have been flawed. For instance, Tesla may go bankrupt if all subsidies including tax deductions are halted, and no other effective climate policy is developed. And supply shortages of essential materials such as lithium and cobalt may reduce required production volumes. Similarly, taking a different direction in public governance will yield different climate policy instrumentations, and different specific technologies may very well follow, as in shifting from electric drives in transport to hydrogen fuel cell drives, as advocated in Japan.

It is possible to distinguish several approaches to structured design covering several levels of abstraction and with several design models (Joore & Brezet, 2015). The strategic level has been explored by Geels and Schot (Geels, 2002; Geels & Schot, 2007). They distinguish sociotechnical landscapes, which are mostly stable, and multi-actor networks of sociotechnical regimes involving market preferences, industries, policy, science, culture and technologies. These sociotechnical regimes evolve more dynamically, inducing slow changes in their sociotechnical landscapes. Within this landscapes and regimes context there are niches for more specific design, not always fitting and welcome in the regimes they are to become part of, and often necessitating that active resistance be overcome. Regime resistance to climate innovation is common when developments go beyond niches, as demonstrated in the climate cases described by Geels (Geels, 2014). The basic reasoning behind leveled approaches in design is laid out by Roozenburg and Eekels (1995). The six levels of considerations distinguished here are a simplification that roughly follows their analysis, with the final reflection step being a possible starting point for iteration of the design process following performance modeling or actual experience. Defining the starting points for design clearly has creative and not directly administrative-technical elements. The future is actively created, even if only partly so, based on the discovery of what is possible and desirable.

With respect to climate policy instrumentation, the core function of reducing emissions may not be a direct part of public policy instruments. It is the niche disruptions by policy instruments, with regime changes and landscape changes following, that create the mostly private processes leading to deep emission reduction, including innovation, in any policy instrumentation design. Emission reduction is more directly linked to the regulation of technologies in Planning & Control, and it is more abstractly incentive-oriented in Institutionalism. Planning & Control regulations also function, however, at some distance from actual dynamics in industries and technologies, as private firms, markets and consumers play a key and independent role in the dynamics of emission reduction. Rebound mechanism will detract from specific reductions while on the other hand induced technology development may induce further emission reductions without further instrumentation, as was the case with feed-in tariffs for solar PV cells. Regional and highly local cultural and institutional elements involved in innovation are not easily shaped using centralized policies (Mancinelli & Mazzanti, 2009). An ambitious policy example for improved regional innovation was attempted in Scotland but seems a complete failure (Brown, Gregson, & Mason, 2016). Central intentions may not materialize and may even break down local capacities.

Deep emission reduction toward 2060 will have to overcome the usual slowness in innovation implementation. The pace of deep innovation is slow in general as documented by Hirooka (2006). His examples show that successful deep innovations do not achieve maximum speed for around 35 years — his "35 years rule." Similarly, but specific for the energy system, Kramer and Haigh (2009) show that new energy sources can grow very fast when volumes are still small. But above a share of a few percent in total energy supply, the speed is reduced, in line with the 35 years rule. There is not one single restraining factor but rather a web of factors, ranging from specific resources required, infrastructure to be designed and implemented, and complex regulations to be adapted, to market control by endangered firms and political pressures resulting in safeguarding of existing production.

When low oil prices, below $50 per barrel, endangered North Sea oil production, the UK government reduced taxes to keep the oil flowing, in parallel to a highly ambitious national climate policy. Regime resistance is all-pervasive.

The central message to take into account in instrumentation design is that there are many layers of more general and more specific considerations, with higher level considerations informing more concrete choices, and generalizations taken from the concrete being helpful in finding solutions for more general problems. An example to clarify this point concerns emissions trading. Tradable emission permit systems are difficult to apply to smaller firms due to high public and private regulatory costs. Mobile sources of emissions such as transport are also difficult to cover. The relationship with biofuels in the mix is very difficult to administer. These problems vanish, however, if measurement shifts from the level of actual emitters to the upstream suppliers of fossil fuels, such as oil and coal companies and natural gas suppliers. The application of a cap-and-trade system can then be broadened to cover all fossil emitters equally. The emission permits are no longer bound to any specific firm, thus reducing options for specific firms to have additional permits supplied to them, which was one contributing factor in oversupply of permits in the EU-ETS. Political vulnerability is then reduced. The emission pricing applied upstream translates to actual emitters through normal market mechanisms. Effectiveness is reduced where such mechanisms are lacking. This may be the case, for example, if the emitter is publicly owned by the same authority that also receives the proceeds of the emission pricing, as is sometimes the case in communist countries at the provincial level. The regulatory effect may then be nil, especially if policy domains are well integrated. Such considerations lead back to regime and landscape aspects of design. A possible solution would be to have ownership and proceeds administered by different public authorities, or to privatize the emitter. Such issues are now to be resolved with respect to provincial electricity production in China.

From political philosophy to strategic and operational instrumentation design

Instrumentation design, as strategic design, may take lessons from design for innovation, as strategic design also requires more than simply technical analysis to improve versions of instruments. The layers of analysis distinguished in the Geels-type design methodology go from the concrete and variable to the malleable to the mostly fixed, in the design process, moving from niche design to related changes in regimes to mostly fixed landscapes (Geels & Schot, 2007). This triple layer system can be used for empirical analysis of how innovation processes work (Foxon, 2011). Such an analysis, however, does not distinguish between "who does what" in a public policy action sense. In a later paper, Geels et al. advocate a climate policy instrumentation approach linked to this triple-level innovation analysis (Geels, Sovacool, Schwanen, & Sorrell, 2017). Their approach takes into account several empirical relations in society, and shifts attention from niche innovation to an integrated approach at the regime level, their "sociotechnical system." There they suggest more specifically a broad range of governmental action: low-carbon innovation policy should focus not only on research and development (R&D) and financial incentives in niches, but also on experimentation, learning, stakeholder involvement, social acceptance, positive discourses and opportunities for new entrants, and substantially using subsidies, thus moving also to regime-level changes. In their approach, stronger alignments are then necessary between innovation policy and sector-specific policy (with respect to electricity, heat, transport and urban planning) to explore the potential of interacting technologies and systems, with elements of landscape changes. Polycentric efforts, in particular, should connect and align scales, actors and responsibilities for increased effectiveness from niche to landscape. Furthermore, accelerated low-carbon transitions also involve actively phasing out existing systems, taking into account the social and distributional consequences of decarbonization. They conclude that in order to create real-world transitions, it is crucial to combine technically feasible, least-cost pathways with innovation processes, business strategies, social acceptance, cultural discourses and political struggles (Geels et al. 2017). The wider scope of empirical analysis of innovation has been translated into a program for broader policy, with concomitant broadened instrumentation as well. This would seem to require not only an unprecedented level of policy control of private activities, but also an unprecedented level of integrated policy itself, going beyond specialized governmental organizations and ministries. Deep innovations could then be stimulated, assuming that all this is possible.

In an earlier introductory paper in a special issue of *Research Policy* (Innovation Studies and Sustainability Transitions), however, Smith, Voß, and Grin (2010) highlight the difficulty of connecting multiple sociotechnical regimes, if only because there are so many of them and they are not well defined. This is the issue that also hampers mainstreaming of the climate problem, and other issues as well, across all policy domains. It is important to work with the different levels involved in innovation, with analogues in policy innovation, as illustrated in Table 3.2. Geels et al. choose an extreme version of the Planning & Control approach. Their framework may be used more modestly, however, to clarify the

Table 3.2 Instrumentation design at levels of landscapes, regimes and niches

Levels in transition	*Levels in Planning & Control climate policy design*	*Levels in Institutionalist climate policy design*
Landscape changes	Not changed directly, possibly some adaptations	Political philosophy with institutional adaptation
Regime/sociotechnical system changes	Induced by niche changes and some for easier niche changes	As required for changed institutions
Niches changes	Technology-specific instrumentation prime • Specific designs • Effective innovations	Technology and market changes resulting mostly autonomously

differences between the two governance approaches for climate policy instrumentation. In Planning & Control climate policy, instrumentation design is similar to niche designs: how to design a new passenger car, taking into account the developments in the regime, in that case the evolving transport system, including the developments in hybrid cars, electric cars and self-driving cars. The limitations of the landscape are taken as a given: the rule systems of society, the main infrastructure and normal economic dynamics. The regime as the sociotechnical system is also given, for the most part, changed by the niche designs and by the adaptations required for their broader success. The sociotechnical system is not changed independently. This order of reasoning is reversed in Institutionalism, which would ask: what key elements need to be changed in the landscape superstructure, and what regime changes are necessary to induce them accordingly? Niche design is not present in the policy domain in this governance approach. All climate-relevant niches will be influenced in their dynamic development, not by direct policy interventions but by the changes in landscape and regimes.

Next to leveled approaches to policy instrumentation for transition design, there also exist seemingly extreme simplifications. One example is entitled "Emissions inevitably approach zero with a 'carbon law'" (Rockström et al., 2017). The carbon law lays out an emission-reduction path for reducing emissions by at least half every decade. The roadmap links long-term requirements for the 2- or 1.5-degree climate goal to short-term targets, to be revised with new narratives every 2 years (p. 1271). Adoption of the law, so goes the reasoning, will require governments and private partners to do what is necessary to implement the law. "A carbon law applies to all sectors and countries at all scales and encourages bold action in the short term" (p. 1269). This simplification seems to beg the question as to how to achieve this, while at the same time specifying large numbers of technology requirements, indicating a Planning & Control approach, with little elaboration on the specific content of instrumentation. A few incompatible instruments are listed: emission taxes, cap-and-trade with a price floor, feed-in tariffs and quotas. This approach to instrumentation seems too narrow and direct, because some of these instruments are contradictory (cap-and-trade and feed-in tariffs) or incompatible (carbon taxes and cap-and-trade). An overall requirement is imposed to start emission pricing at $50 per tonne CO_2 and increase (by around $17 per year) to $400 by midcentury. This core requirement goes against Geels et al., who see the encompassing emission tax as politically unfeasible. They prefer domain-specific subsidies funded by other taxes (Geels et al., 2017). The *carbon law* could prove to be counterproductive. The good intention, to tackle everything with all means at the same time, may turn against itself. It may become clear over time that little has happened due to complexity that was not strategically resolved at the outset or early in the design process. Geels et al. (2017) may also overcharge the political and administrative system, but in a much more structured way.

The two approaches developed here are more tedious and may be less likely to elicit broad public enthusiasm along the lines of "yes, we can!" They entail broader reasoning toward developing consistent sets of policy instrumentation. In the long run this might be more effective in either of the two approaches developed and linked to the two basic directions in political philosophy that have been present for several centuries (see Chapter 2). Within each direction, clear strategies should lead to coherent sets of operational instruments that are fit for the deep reduction task aimed at in the good intentions. The distinction of levels by Geels and Schot (2007) remains highly relevant as a conceptual background in instrumentation design.

Systems analysis for reducing fossil CO_2 emissions

Carbon flows and CO_2 emissions

Physical systems analysis forms the basis for climate policy and its instrumentation design: CO_2 emissions are physical entities resulting from carbon oxidation. The inflow of carbon is to be reduced or the non-emission outflow is to be increased, as permanent storage. The buildup of stocks of carbon in the economy may delay the outflow, as in durable applications such as polymer building materials and plastics in managed landfills. Volumes are diverse and small, and are difficult to measure and control over decades. So this seems a not-so-relevant issue for long-term deep climate policy. Bringing less fossil fuel into the economy is what ultimately counts. The inflow of fossil carbon is fully predictive for the outflow as CO_2 emissions, reduced only by geological storage of carbon, as in carbon capture and storage (CCS). At a global level the system is simplest; see Figure 3.1. Some complexity arises as CO_2 can also be removed from the air by biota and industrially by direct air capture (DAC), including non-fossil CO_2; some emissions also occur at incineration from biota mixed with fossil CO_2. In general, the quantity and specific sources of flows from air and to air are difficult to measure. The flows from geology and to geology are, conversely, easy to measure. A global government would have an easy implementation task.

More complexity arises from the fact that climate policy is a national responsibility, thereby including more decentralized public bodies as well. At a national level, imports and exports of carbon complicate the picture substantially, with high

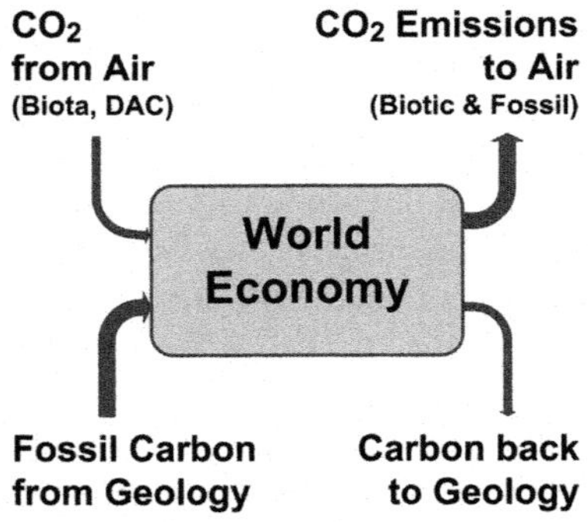

Figure 3.1 Carbon flows through the world economy and resulting CO_2 emissions.

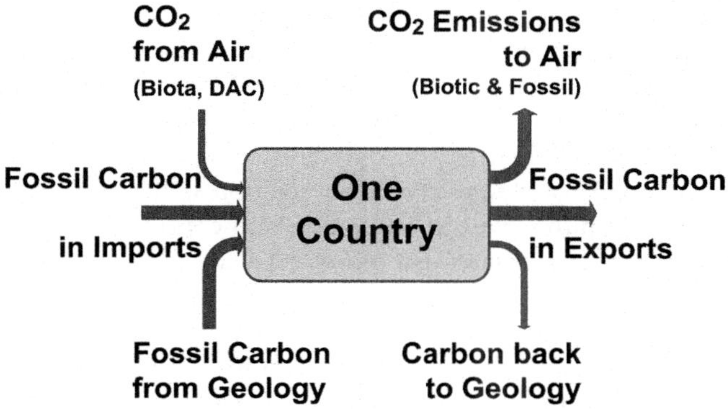

Figure 3.2 Carbon flows through a national economy and resulting CO_2 emissions.

carbon concentrations in fossil energy flows and in organic base chemicals as for polymers. But virtually all other products contain some carbon, as in plastics, glues, coatings and solvents. Globalized production leads to a substantial international share in national carbon flows; see Figure 3.2. Emissions have their main background in energy use of fossil fuels, including from plastics in waste incineration.

At a somewhat deeper level of detail sectors can be distinguished. The sectors most commonly used in climate policy – energy, industry, transport and buildings – hide an extreme diversity. Input-output tables tend to increasing detail, with around 100 sectors per country now connected internationally. But diversity per sector is large, even if the number of production sectors distinguished increases to on the order of 500. The UN has a standard sector classification for this purpose. Linking 500 sectors per country, with a quarter million connections between sector cells, to all 100 main countries in the world equates to around two and a half billion connections to specify, most of them not zero either in economic terms or in carbon terms. But even at that level of detail, the firms in one such detailed sector are highly diverse and their products even more, and classifications become increasingly complex. A mobile phone is also a radio, a TV and a data storage device, and mobile phones vary widely, changing rapidly over time. Going beyond the 500 sectors may deepen insight. But the number of firms is globally in the order of a quarter billion, and the number of private households is several billions, each of them making partially independent decisions on what to buy and how to process their inputs into different intermediate and final outputs, all evolving in time. It certainly is not possible to cover that detail of carbon flows and CO_2 emissions in any real way. Somehow complexity is to be reduced to make climate policy a feasible option, covering 95%, 99% or 99.9% of all carbon flows and emissions.

Policy instrumentation from a systems point of view

There are two solutions to make the socioeconomic complexity manageable for climate policy. One is to take the national full systems-level boundaries as a starting point. Inflows, as primary fossil fuel production and imports, and outflows, as exports and CCS, can be linked to policy instruments, be they volume-type in Planning & Control or pricing-type in Institutionalism. This is referred to as the volume or the price approach by Christina Hood (2011). Fossil CO_2 emissions are measured as the difference between these national fossil inflows and economic outflows. The other solution is to focus on main emitters and groups of emitters within a country, and control their emissions. For each controlled emitter, measurement is indirect again, based on its fossil carbon input minus its CCS, as direct emission measurement is mostly not possible. Most emissions are small scale and decentralized now, as in transport, building heating, and many industries, where indirect measurement is costly and can hardly be applied effectively. So some form of aggregation of emitters is required for covering the full total of emissions.

The millions of transport vehicles in a country, for example, together have a share of around a third in total emissions. They can be aggregated in different ways. One is as to their fuel suppliers, with the carbon content of fuel for measuring emissions. This direct emission measurement allows for application of direct emission-reducing instruments, be it by one overall cap also covering other sectors, one transport cap, or differentiated caps per transport type, or by one emission tax also covering all other sectors, or a specific tax covering all transport, or a tax differentiated per transport type. The cap must induce reduced carbon use in fuels, in principle through a basic price mechanism: a higher price. If the cap is grandfathered, with fuel marketing certificates based on last year's supply, the higher price goes to the fuel suppliers, creating windfall profits as short-term demand is inelastic. Or the fuel certificates are auctioned, that profit then accruing to governments. The EU-ETS has a separate tradable cap to cover aviation only. With a tax at similar levels, similar proceeds accrue to governments directly. It seems a needless complexity to differentiate the emission tax between users of fuel. All such options have in common that they link to emissions directly, though not all equally.

The other option for aggregation in transport is to link emissions of vehicles to their producers: car, van, bus and truck producers for road transport, but also the producers or traders in tractors and other agricultural and garden machines, and many more mobile apparatus, like cranes, bulldozers, etc. Also shipping, rail and aviation are then to be covered through their suppliers. Current fleet standards follow this approach, covering person cars, vans and small trucks. The instrument link is to emissions only indirectly, by proxy. Fleet standards link only loosely to system-level transport emissions, as the volume of the fleet is not regulated, nor are driving style, speed limits, modal shifts and more factors determining total emissions of each fleet and the sum of fleets. It is difficult to set fleet standards for the other more diverse transport vehicles, and it is difficult to control all decision makers relevant for ultimate total emissions. It is the outcome at full system level that counts, added over all transport fleets and ultimately all sectors.

Classification of instruments for climate policy

Climate policy instruments: what they specify is what they do

How can instruments be specified for the goal of instrumentation design? There is no unique or best specification available as the functions of classification and categorization differ. *Legal status* leads to specifications such as covenants, operating permits and binding standards and regulations, linked to administrative organizations and sanction types. The *target of regulation* may be *result*, *technology*, or *behavior* oriented, such as in emission permits, fleet standards and speed limits for road vehicles. Or categorization may differentiate among *objects regulated*, such as in instruments for products, installations, or behavior. The *scope* of application may concern an individual person, a firm, a group of persons or firms, a particular sector or a product-technology class. Fleet standards, for example, have a broad scope for a limited technology domain, covering all vehicles in a particular class such as person cars and vans, by one producer, and one target, the average emission of the fleet sold driving for 1 year as measured in a standard test drive. See Huppes and Simonis (2009) for a survey of such classification principles in the environmental policy domain. Instruments must always have a specified legal status, a defined object and a direct target — that is what the instrument sets in motion. A tax on some energy flow reduces that energy flow and a cap or tax on some emission reduces that emission.

The primary characteristic of the categorization of instruments used here is their instrumentality: what they target and set in motion is what they do, regardless of intentions, broader goals, indirect effects and legal status. Emission pricing, from taxes to fines, incentivizes emission reduction; an emission standard allows for actions that do not exceed the standard; a continental transmission grid opens the option of continental electricity transport; product information allows for better-informed choices; a technology prescription limits the technologies that may be applied, such as *no coal-fired power plants, no nuclear energy* or *zero-energy buildings*; and a subsidy on roof solar PV stimulates roof solar PV. Justifications for such measures are certainly more complex and abstract, with goals and values involved, complex mechanisms set in motion and with practical restrictions guiding choices. But as instruments, as public policy measures, they are simply what they are, regardless of intentions, including good intentions or hidden private interests. The reasoning linking instruments to emission reduction is direct for emission pricing, emission permits and emission standards, or indirect, by proxy. There is a difference in this respect between an emission tax and an emission permit. The emission tax creates costs directly. The permit requirement creates costs indirectly, and differently for different firms. If emission permits are made tradable they become emission rights, still indirectly creating a price, but then a more uniform one. Firms try to sell the rights they have by reducing emissions, or thus have to buy them, as for expanding production with higher emissions. An adequate market can create a uniform price for the emission rights. The emission tax creates this price directly. The expected future price guides R&D and investment decisions.

The intention of an instrument may be CO_2 emission reduction, such as in subsidizing solar PV with feed-in tariffs. What the instrument does, however, is make solar PV more attractive for its delivery of electricity to the grid. Indirectly this PV electricity may then be more attractive for direct use by the producer, depending on market circumstances and storage options. Whether the PV cells installed and delivering to the grid will, indeed, lead to present or future emission reduction of some volume depends on the causal mechanisms set in motion by this instrument only indirectly, different in different circumstances, and depending on further future actions of other both public and private actors. In countries with a small share of fossil fuels in electricity production, solar PV electricity to the grid will not press out much fossil emissions but will mainly press out non-fossil electricity production. Views and models may differ in this respect taking into account different mechanisms under different assumptions. For example, fragmenting the electricity market, as feed-in tariffs do, reduces market mechanisms and so reduces the attractiveness of heat pumps and electric cars; in time this might even lead overall to higher emissions rather than lower ones. One main assumption in such an analysis of indirect effects is often that such negative indirect effects may be prevented by applying additional policy instruments. Of course, additional instruments may be applied, and in very different ways, such as higher taxes on natural gas and higher subsidies on electric cars, or encompassing emission taxes, or a cap-and-trade system. A complete instrument mix must somehow cover such indirect effects if they would detract from the overall emission reduction goal. This could then easily lead to a substantial number of intricately related instruments. Broader strategic design considerations should serve as a check on such a proliferation of instruments intended to make the core climate policy instrument successful. Simplicity, including sparseness, is one such design principle for overall instrumentation, as described later in this chapter. In designing instrumentation there must be an overview of instruments, such as in the typology developed below: how to choose and detail specific instruments suited to the development of relevant overall instrumentation, with a method and narrative. First is a survey of two other approaches to overall instrumentation.

Other approaches to climate policy instrumentation

There are many studies that analyze specific, single instruments for climate policy which then also provide options for their detailed improvement. Studies with a view to overall instrumentation are, however, sparse. They require specific views on governance, an issue not well appreciated in the very diverse political contexts of climate policy. Two major exceptions are noteworthy: Grubb, Hourcade, and Neuhoff (2014) and Hood (2011). Both have an overall view of instrumentation. Grubb, Hourcade, and Neuhoff go beyond direct welfare considerations and look into what is needed practically for the deep, longer-term transformation required for achieving a near-zero emission society. Hood sticks to an economists' welfare approach, with a focus on either quantitative restrictions or financial incentives. Other approaches, such as by Geels et al. (2017) and Rockström et al. (2017), list very many instruments but do not systematically develop an overall instrumentation. They sketch political and strategic approaches to any instrumentation but do not develop a typology or classification from a governance, regulatory or effectiveness point of view.

Grubb et al. start with ordering policies in three time frames. ***Satisficing***, following Simon's satisfying (Simon, 1956) and other behavioral economists (Cyert & March, 1963), creates short-term improvements, possibly in series, which is Lindblom's *muddling through* (Lindblom, 1959). ***Optimizing***, the core of much of economics and especially applied welfare theory, improves on satisficing, but works in the medium term, only. ***Transformation*** is required for deep reductions in long-term climate policy, beyond the domains of satisficing and optimizing. These time frames link to three pillars for instrumentation. *PILLAR I –Standards and engagement for smarter choices* (Grubb et al., 2014, p. 79ff) links to satisficing, to what can practically be realized now. What improvements are possible in the main technology domains now, such as in low-emission power, smarter buildings, cleaner production, transforming vehicles and fuels and smarter grid systems? The core instrument of *PILLAR I* is adapting standards and regulations, making them more stringent. In "barriers to and drivers of change" (p. 135ff), they indicate the options and limitations of such shorter-term policy instruments. *PILLAR II – Markets and prices for cleaner products and processes* (p. 203ff) aims to improve efficiency and optimality. Resolving market imperfections through pricing measures is part of optimization. Reducing fossil energy subsidies is a main first step (p. 211). Specific product taxes could be transformed to more generic pricing for pollution (p. 215). Two options for emission-pricing are compared based on their efficiency, favoring emission taxes in the short term and cap-and-trade systems in the long term (p. 225). How such instruments are to be implemented is left open. *PILLAR III – Strategic investment for innovation and infrastructure* (p. 311ff) goes into the deep societal changes required not just in the energy system but for lasting deep emissions reduction, all of which also require innovation with public funding for their implementation: "Pushing further, pulling deeper: bridging the technology valley of death" (p. 315ff). The entire innovation chain has to be covered, using technology push and demand pull. Instruments include public funding of R&D and Demonstration, with subsidies and pricing schemes for strategic deployment. These create learning curves to overcome lock-in barriers and require optimization of infrastructure, complemented by technology prescriptions. Grubb et al. indicate the transformation pathway for four main sectors: energy, transport, industry and the built environment, all in a technology-specific and market-specific vein. The transformation of the European electricity system is extensively described (p. 372ff), for example, but mainly in a technical sense. They have no systematic treatment of the public policy instruments involved in long-term transitions, but in their examples they link most directly to the Planning & Control mode of climate and energy governance.

Hood, for IEA/OECD, takes an economists' market view of instrumentation (Hood, 2011). Emissions can be capped with permits, made tradable in cap-and-trade systems for more flexibility and efficiency; they can be priced directly, as in an emission tax; or instead of caps or prices, they may have other forms of more technical prescriptions. Other instruments may be added to each basic option, always with cost-effectiveness and dynamic efficiency in mind (see Hood, 2011, p. 38ff). But cap-and-trade, she argues, is often combined with other domain-specific instruments, creating serious problems of overlap and misalignment. Varying the cap to keep the price at relevant levels creates a hybrid system, partly volume and partly price based. The emission tax then is simpler. It combines well with energy efficiency measures and support for renewables for the purpose of reducing overall cost. "As in a trading scheme, the primary justification for renewables support is long-term cost reductions" (Hood, 2011, p. 42), including creating learning curves by inducing and implementing innovations. If an encompassing emission tax cannot be implemented, the reference for more technology-specific policy is to specify and enforce what could be expected to happen under an assumed ideal emission tax and with perfect markets. Specific instruments are to mimic these outcomes, then involving quantity measures such as renewable energy and clean energy standards, or pricing instruments such as *production tax credits* for particular forms of renewable electricity. Incentives for improving the reduced dynamics resulting from technology-specific instruments are to be added. There is no generic design solution for detrimental effects of contradicting instruments, but Hood has strong, somewhat optimistic advice: "The overall conclusion is that interactions can be managed as long as the focus is on functionality of the whole mix: coordination of the various targets, and the detailed design of instruments taking account of interactions" (2011, p. 43).

Moving toward ideal emission pricing and toward ideal markets is her first best instrumentation option. Only if the generic emission tax remains elusive, such as for political reasons, will the substantially more costly technology-specific option have to be developed, in the stay-close-to-the-optimum approach. It is as a second if not third best option, which will lead to ever more complex instrumentation. Hood thus remains mainly in the Pillar II domain of Grubb et al.

Policy instruments classified: six main types

The core of any instrument is what it sets in motion via its working mechanism if effectively applied, and not its intention. One layer of instruments is focused at regulating private parties, producers and consumers, in setting standards as prescriptions and prohibitions, subsidies and taxes to guide investment, production, purchasing and use choices, and information to allow for such choices in a more informed way. Emission standards per kilowatt hour on fossil power stations phase out power stations above the standard. An emission tax on fossil fuel power stations raises their cost of production, different for different fossil fuels. Such direct climate policy instruments can reduce emissions directly, with other mechanisms following. A proxy variant, such as an energy tax on electricity of the same total amount as the emission tax, will have a more limited emission-reducing effect, as there is no differential effect on highest emitting electricity production. It taxes non-fossil electricity equally as well.

Technology-specific instruments can be binding, or incentivizing, or options creating, by spreading knowledge on emission-reducing options. These three types of instruments influence private behavior by firms and consumers. Pricing schemes to reduce emissions are less binding than prohibitions and prescriptions but create incentives. Heavy taxes can come close to prohibitions. Novel information leaves decisions fully to the ones receiving it, opening up options only for them. The sequence of reduced bindingness can be followed further. Creating novel information is one step further, as by publicly funded research and development. That information may next be used in novel marketable options. The outcomes are not binding on any private actor as through regulations and financial incentives, they create new options. Of course, if the option exists, it can be incentivized or forced into the market by prescription or by prohibitions of its competing options. Similar types of option creation relate to publicly provided infrastructure. A road, an internet cable, or a harbor may be used by a decision maker, or not. Again, when developed, further rules of the more specific types regarding private actors can be added, as in opening the new road connection or lane only to electric cars or those with several passengers.

The most generic route to climate policy is adapting the rules of the game, the institutions, roughly coinciding with Frank Geels's landscape level (2002, 2012). Geels sees the landscape as mainly stable, with incidental infringements. Institutions of industrial states have some core stable elements in their setup. But they have developed first and are changed continuously however. Rules on ownership, as on liability, markets and taxes, are in continuous flux to adapt to new developments. Rules on intellectual property rights and its protection change nearly continuously. Basic infrastructure can develop fast, based on standardization, with mobile phones and the internet as recent examples. These may be mainly public or mainly private, depending on details in rules. When private, additional more specific market rules will apply, as on the number of competitors in mobile phone operators. Net neutrality at the now mostly privatized internet may stay, or not, with reasoning based on societal advantages, driven by specific interests. Telephone and electricity have long been public monopolies, for over a century, but have substantially been privatized and globalized, resulting in iPhones and decentralized solar PV. This is the generic level at which climate policy instrumentation starts, in the Institutionalist mode of governance.

The six types of instruments distinguished are summarized in Table 3.3. They are general in setup but can be filled in for climate policy, specifically.

So, again, what are policy instruments? The first step is to work with the layers of considerations in design, and see how climate policy can link in. Using this design background and following a basic distinction as to how current market-based societies operate, the bewildering number of practically possible instruments may first be reduced to two main types: public tasks and operations are separated from regulating specific private market operations. This

Table 3.3 Instrument types from general, incentivizing and enabling to specific and binding

Instrument types
A. Public tasks instruments
1. Institutional framework: climate liability, real-time electricity markets, emission taxes
2. Public infrastructure, spatial planning and standardization
3. Publicly funded research: R&D, Demonstration (RD&D) and Market Development
B. Regulatory instruments regarding private actors
4. Informational instruments: on technologies, products and behavior
5. Financial instruments: taxes and subsidies, including tax deductions and differentiations
6. Standards and regulations: prescriptions and prohibitions

distinction relates to political philosophy and may shift over time among various socio-democratic, liberal and neoliberal variants, and in liberal societies this distinction operates very differently than in communist societies with a greatly reduced private sector. Private actions are guided by how markets are created, mainly in the institutional superstructure of society, its landscape and regime in design terminology. This connects to the first distinction between (A) *public task instruments* and (B) *regulatory instruments regarding private actors.* Public tasks first relate to landscape development as the most general level, here corresponding to the institutional framework of society, instruments of type 1. They include liability rules, intellectual property rights and the tax structure. These changes are shaped within the overall the constitutional framework providing how decisions in society are made and how conflicts are resolved. Type 2, roughly corresponding to Geels's regime level, can be described in somewhat more detail as referring to spatial planning, standardization, and the supply of public goods such as health care, education, fundamental research and infrastructure. As deep emission reductions will require deep innovations, for climate policy there is a subcategory on innovation as publicly funded research, R&D and Demonstration, public task type 3. As with type 2, it creates options for use by private actors. These instruments are not binding. For type 2 and type 3 instruments the boundary with private actions may be defined differently, related to political philosophy and governance views and to ideas on how to treat urgency in climate policy. If more technical developments are assumedly required soon, like the hybrid-electric cars and other vehicles, fully electric vehicles and hydrogen fuel cell vehicles, private development may fall short of what is deemed required. Then public development or publicly supported development may become a public task. In the car example, it was not public development but public subsidies that created incentives for market development, with public research focused at more specific issues, such as novel battery development.

In regulating technologies and the behaviors of private actors, the bindingness of instruments is a basis for further differentiation. Least binding is information provision, either setting the way information is to be provided, or providing information directly to the public so that the audiences involved can decide themselves what to do; that is type 4. Type 5 is financial regulations, which are flexible in setting the size of the shifts in the desired direction for regulatees. A normal tax on aviation fuel will decrease flight volumes overall, but does not prohibit a round-the-world trip. Type 6 comprises more binary regulations that prescribe technically defined limits to be met, found in general standards or in individual permit regulations. However, each of these instruments can be filled in a more or less general and encompassing way. In regulating fossil power installations, CO_2 permits per kilowatt hour for any fossil electricity-producing firm are more encompassing and require fewer regulations than permits made individually per installation. These can be softened and made more incentivizing by making the permits tradable, still to be applied to individual installations.

Public task instruments exemplified for climate policy

Type 1, ***Institutions,*** primarily create incentives and options. They include the use of tort law for holding emitters liable for damages, an emission tax for internalization of external effects, adapting the electricity market to better deal with intermittent supply, and clearer ownership of underground heat and cold storage. Legal liability as created by tort law can be developed and used in ways similar to the imposition of liability for other negative environmental impacts on third parties. This is a prime option according to Faure and Peeters (2011). However, this option is less suited to global warming and ocean acidification for several reasons, individually, and the more so combined. The number of emitters is very large, literally billions. Their emissions after mixing globally are all linked together to negative effects on very large numbers of persons and organizations, also in the billions. Emissions effects have substantial time delays, with past, current and future emitters together creating effects of different natures that will manifest much later, some of them well preventable. And last but not least, the future effects to be expected are not fixed but are an array of possible effects, the most serious ones with a low and barely known probability of occurring. A *fat-tailed* chance-effects structure has been described (Weitzman, 2009, 2011). These low-chance effects are catastrophic and beyond measurement in economic terms. Holding polluters accountable to victims is simply not possible in a private law sense. Even when a single explosion directly killed and injured thousands of people, as happened in and around the Union Carbide chemicals factory in Bhopal, interrelated national and international institutions prevented effective liability to be implemented through tort law and criminal law. There also exist theoretical reasons for not making climate change a subject of ownership and tort law. There are inalienable entities not open to trading (Calabresi & Melamed, 1972). Humans, most beaches and legal protection are not tradable; there exist inalienable rights. The right to climate stability may also be considered inalienable, hence not subject to trading between parties. This would go against the idea of tradable emission permits, and also against free individual emission permits and operating permits allowing emissions.

Following in the line of liability reasoning, but then not with private legal means, is the internalization of external climate effects applying the "polluter pays principle" as specified in the guiding principles of OECD (1972). A generic emission tax is the prime option, following early ideas of Pigou on how to correct market deficiencies (Pigou, 1920/1932), adapted for more practical use in policy by Baumol and Oates (1972, 1971). Setting yearly emission permits for major emitters to the desired total level is not an Institutionalist instrument, also not when making the permits tradable and even not if auctioning them. In an idealized economic sense of "pricing" there may be a substantial overlap in the working of an emission tax and a tradable permits cap. But in a political philosophical and legal theoretical sense they are miles apart. There are also deep differences in an administrative sense. One is the complexity of establishing emissions administratively to link activities to individual permits. All cap systems leave out CO_2 emissions of smaller and mobile sources. Thus they are incomplete. In trading, they also tend to create vast money flows in their trading to avoid other tax payments. Tax measures must be added to avoid evasion, which undermines simplicity. Such mechanisms are nearly unavoidable in mixed systems of direct regulation and derived pricing options. In an economic sense there is also the problem of price instability of the emission permits, quite inherent due to inelastic supply and short-term inelastic demand. In a political sense there are also differences, as firms under financial stress may lobby for free permits, leading to oversupply and structurally low prices. More emitter-specific regulations, such as in permits, are politically more vulnerable, even if made tradable. The administrative link to specific emitters, as in the tradable permit, is not part of the upstream emission tax in Institutionalism, as that link comes about through normal market mechanisms. The different governance nature of the emission tax is apparent from its alternative name: the carbon deposit, as an example of a substance deposit (Huppes, 1988). The deposit is paid on fossil carbon entering the country by import and primary production and is refunded upon export and transfer back to geological storage. Those who lose carbon in between, that is, emit it, lose the deposit. That is your own responsibility.

There are few main options for substantially expanding non-fossil energy production. Renewables such as wind and solar electricity can be expanded to a near limitless volume, whereas nuclear energy has safety impediments. The intermittency of most renewables, their limited dispatchability and their substantially decentralized nature make current electricity markets inapt. They need to be adapted. An Institutionalist market design links private producers, small scale and large scale, primary and secondary, to private users, large and small, in one real-time electricity market, which has to be created. This requires public ownership of inherently monopolistic transmission and distribution grids. This approach goes strongly against the current neoliberal strand of political philosophy of continuing privatization of electricity grids, such as is still the case in Australia and the United States, as well as, less so, in Europe. Creating real-time electricity markets is made possible by novel IT-based developments, reacting directly on price changes (Hogan, 2014; Wang et al., 2015; Yi et al., 2013). Enlarging markets can also even out differences between varying supply and demand, also enabled by new technical developments for low-priced long-distance transport using high-voltage direct current (HVDC), in a supergrid. The EU Unbundling Directive (European Commission, 2009) contains more detailed requirements on allowable grid ownership in EU member states (European Commission, 2013). It reduces the monopolistic tendencies of large producers also owning the transmission grid, but it still allows private ownership of transmission grids and distribution grids. An example is the UK National Grid, a private international company also owning production facilities, albeit in other countries. The full and clear separation of market-based, private activities and publicly owned, natural monopolies forms the basis for Institutionalist redesign. The supergrid, the transmission grid and the distribution grid may have separate ownership linked to the corresponding administrative levels. The Chinese supergrid, which has substantially been built, is not fully used to purpose as the provincial transmission grid owners also own the provincial production facilities and do not want their prices depressed by periods of cheap renewables supply from other provinces. The separation of ownership of production, supergrid, transmission and distribution is fundamental in the Institutionalist governance approach, but not in the Planning & Control approach, which focuses on subsidized and regulated renewables and controlled nuclear expansion. With limited development of real-time markets, limited storage capacity also will develop autonomously and limited user response can arise. Security of supply then requires subsidies to fossil energy producers as costly backup, with emissions. This is the current Planning & Control line, which, however, can be improved upon.

A final subject, at the boundary of climate policy, is intellectual property, increasingly protected in patent systems, both by better protective procedures and by expanding the domains and duration of protection. Such developments may delay innovations and their market introduction (Boldrin & Levine, 2008; Jaffe & Lerner, 2004/2011, Stiglitz, 2014). Additionally, patent numbers in renewable energy, storage systems and zero emission technologies are exploding. Countering this monopolistic tendency certainly is an issue in an Institutionalist approach, but not really part of climate policy instrumentation.

Type 2, ***Public infrastructure, spatial planning and standardization,*** covers subjects of an inherently centralized nature. ***Public infrastructure*** involves roads, railroads, electricity grids, harbors to connect to cities, industries and energy and materials sources. These are closely related to spatial planning. Developments may be left to private parties but then entail private monopolies. This is not acceptable in Institutionalism and is subject of public–private negotiations in Planning & Control, also regarding climate policy. As electricity and possibly hydrogen and maybe metal oxides will play a key role in final energy use, not only novel energy market designs are required but also new public infrastructure is needed to create the options for their efficient use. The planning of a supergrid connecting transmission grids is a public task, as is building the infrastructure. The European supergrid could also connect to North African solar production. A similar system is starting to be developed in East Asia, linking China, Russia, Mongolia, South Korea and

Japan, and yet another is developing around the southwestern regions of China, including Pakistan. Regulated privatization of infrastructure may follow public provision, as was the case with the development of the internet. Mazzucato (2013) advocates expansion of this public investment role, taking a position close to that of strong public Planning & Control. Mingardi (2015) criticizes her as the boundary between public and private domains becomes blurred. From the Institutionalist perspective privatization of public infrastructure becomes an option only if there is substantial competition from other technologies, such as when competitive mobile phone companies made their appearance, which allowed for the privatization of most fixed-line telephone companies. In this non-climate example, which provides an interesting analogue, privatization of the data transmission lines by monopolistic companies, including fiber optics and ADSL, may have occurred too soon in this governance view. Competition from satellites does not offer the same quality. Low-orbit satellites, high-flying solar planes and 5G aerial transmission might create enough competition in data transmission for non-monopolistic privatization of the internet, if combined ownership could be prevented. In electricity transport such a development cannot be envisaged yet. In the Planning & Control approach the focus would be on getting private partners to create public infrastructure in public–private partnerships, with accompanying detailed regulation to avoid misuse of their position and, conversely, public guarantees of private profitability. Especially large companies would gain substantial public influence, going against Institutionalist constitutional views on clear separation of public and private domains. Such mixed developments might well entail stagnation as large organizations will tend to protect their established interests (Acemoglu & Robinson, 2012).

Spatial planning is generally considered an integrative public role. Large-scale solar cell systems do not fit in urban surroundings. The planning of supergrids and hydrogen storage systems may constitute a substantial input in spatial plans. However, the role of climate policy remains limited in overall spatial planning.

Standardization of products at the national level, as well as through the International Organization for Standardization (ISO), becomes a subject of public policy only to prevent monopoly creation by private parties or to overcome standardization being delayed by opposed private interests. Climate policy can then make an entry, as has already substantially been done in the electricity domain for linking transmission grids between different owners. Uniform or compatible plug standards for charging car batteries may be an example, allowing for competition between electricity providers along the road, and likewise for hydrogen supply. In Institutionalism the pure public role might be stronger than in Planning & Control, oriented more toward public–private partnerships also developing spatial plans and infrastructure.

Type 3, ***Research, R&D and Demonstration,*** forms the gateway to more concrete, private economic activities. The main share of research expenditure in developed countries comes from the private sector. Private research is intended to fit into business plans having a short-term component, to attract risk-averse investors not focused on the uncertain long term. Research having a long-term horizon and involving high uncertainties is a public task in Institutionalism. Such research is mostly fundamental research that adds to usable knowledge and then is not subject to temporary monopoly creation through patents. Development and demonstration are only relevant to overcome lock-ins resulting from monopolization, now often based on the currently expanding patent system. The learning curves created by public investment for solar PV and wind turbines would not have constituted a public climate policy task as they do in Planning & Control, since no long-term basic research issues were involved. Making solar and wind turbines competitive is realized by the incentives of institutional redesign and the supply of relevant infrastructure. Remember that the steam engine that changed the world was not the product of a government innovation program, nor were the combustion engine, automobile, airplane and transistor, nor the oil and gas revolutions. The development of the internet was a sideline product of military requirements for safe and indestructible information exchange, a public task in Institutionalism due to the public nature of military protection and next to the monopolistic nature of that infrastructure.

In Planning & Control, conversely, R&D and Demonstration (RD&D) are essential to instrumentation for generating climate-relevant innovation in renewable energy and also in the full production and consumption chains involved, from transport to heating in buildings and highly diverse in industry. Publicly supported market introduction is required for specific technologies as broad market incentives are lacking. The instrument of public co-funding of RD&D then is to cover all of what is necessary for developing a low-emission society, which includes virtually all low-emission technologies for which development is still required. Longer term research has less emphasis in Planning & Control, as shorter term valorization gives priority to Demonstration projects.

Instruments regulating private actors exemplified

Regulating private actors involves three types of instruments: type 4 — information, type 5 — financial incentives and type 6 — standards and regulations, with increasing stringency.

Type 4, ***Informational instruments***, leave decisions to actors, without or with only limited normative pressure; they include product-labelling schemes and information on available technologies such as in describing best available technology (BAT). Information schemes may actively induce innovation by stimulating competition in some domains, such as the Japanese Top Runner program (Hamamoto, 2011; Nishitani & Itoh, 2016). The public role under Institutionalism is in some standardization of information, and in standards for the supply of industry-wide data, a task of type 2 instruments, which do not supply information on specific products and services.

Type 5, ***Financial instruments***, may exert a stronger influence by changing relative price levels, as in subsidies for renewable electricity production and for electric vehicles, and taxes on gasoline and on heavily emitting coal-fired power stations. Next to specific direct effects, subsidies on non-fossil fuels increase the total energy supply while taxes on fossil fuel use, including reduced subsidies, reduce the total supply; see Hood (2011) for examples. In Institutionalism technology-specific financial instruments would not be used as all are covered by rising emission taxes and indirectly by greater access to the electricity market.

Type 6, ***Standards and regulations***, are the most binding instruments and include prescriptions and prohibitions, such as emission standards for power stations and speed limits for private cars. They continue to be the workhorse of environmental policy, and some may be required in Institutionalist climate policy as well, especially for non-CO_2 emissions where emission measurement is generally not reliable enough for taxation purposes. The direct instrument of performance specification, as in applying increasingly stringent emission standards, allows for more technology dynamics than technology specification, a proxy instrument. Even restrictive regulations may induce some innovation, as through prime-mover mechanisms evading restrictions, or just by forcing change in firms; see Porter and Van der Linde (1995) and the more recent literature review by Ambec, Cohen, Elgie, and Lanoie (2013). Institutionalism does not use this instrument type for reducing CO_2 emissions, but may do so for methane and nitrous oxide, where emission measurement required for emission pricing is hardly possible.

How to delimit the domain of climate policy instruments

Climate policy instruments can be defined functionally as any instrument that directly or indirectly reduces emissions. The range then becomes enormous, from contraception and abortion advice and options, to reduced retirement-age rules and longer holidays, to reduced income generation and spending. Mainstreaming climate policy in all other policy domains would ultimately have this consequence. Somehow a more focused domain restriction is necessary, if only to make climate policy workable. The other extreme is where instruments directly refer only to CO_2 and other global warming emissions. These are clearly climate policy instruments. Pure fleet standards and emission taxes are examples. But what if instruments have an emission-reducing effect, such as air quality measures in cities that reduce or ban combustion vehicles? In all low-emission scenarios of the IPCC and IEA, energy efficiency plays a pivotal role. Are energy efficiency instruments climate policy instruments? Is a shift from more road building to congestion pricing a climate policy instrument? Is the obligatory introduction of smart meters climate policy? Is subsidizing floor-heating technology climate policy? Almost all public regulations have emission effects, and when they lead to emission reduction they may be seen as climate policy instruments, and conversely as anti-climate policy instruments. How to delimit the domain of climate policy instruments?

A first step in demarcation is identifying the strength of the link between the instrument and the climate effect. Any public policy instrument has climate effects, but some have more than others. The strength of the link relates to the importance of the non-climate results envisaged. Some measure of such importance is required to make the comparison between instruments according to how much, relatively, they contribute to emission reduction. The importance of congestion pricing is often expressed in terms of the reduction of the cost of congestion. That process may also be linked to emission reduction, allowing the instrument to be expressed in terms of emission reduction per unit of value created. That a distinction can be made between not-climate-policy-instruments and also-climate-policy-instruments is the assumption here. If so, then other instruments with a still higher emission reduction per unit of value creation certainly are instruments of the also-climate-policy type. If the emission reduction per unit of other value creation is really high, at some point the instrument is definitely a climate policy instrument, with a contribution to the other policy goal. And if an instrument reduces emission at a net cost, it is certainly a climate policy instrument. And if some instrument creates a value but increases CO_2 emissions it is certainly ***not*** a climate policy instrument. This reasoning has a strict welfare background, indicating more precise analysis of costs and benefits of specific policy measures.

When a climate policy instrument actively contributes to other welfare-increasing policy goals, it creates a non-climate co-benefit. Reduction of combustion drives as a consequence of some climate policy instrument also reduces other air pollution, subject of health policy instruments. In Planning & Control it is a co-benefit. In Institutionalism it is not. Health effects of air pollution may be reduced due to climate policy. That is one factor among many determining how, and how stringently environmental policy for health reasons is to be developed. Inducing CCS in climate policy, as by an obligatory share in fossil-based electricity production, may well increase non-climate emissions as total fossil fuel energy use goes up, a co-burden. Is that a reason to avoid CCS? In Planning & Control it may well be, as the overall value of CCS policy is reduced. In Institutionalism it is not. CCS may be a necessary part of climate policy while in parallel adequate environmental health policies are to be developed. If indeed such policies would raise the price of CCS so as to effectively reduce it, so be it — and then adapt climate policy accordingly.

Mainstreaming

Reckoning with climate effects of non-climate policies is called mainstreaming of climate policy in these other policy domains. Reckoning in climate policy with its environmental health effects, then, is mainstreaming of health policy in climate policy. Doing so in both directions is a form of policy integration. When more than two policy domains are involved, the analysis will soon become more opaque. The subject of policy integration is treated in some detail in the last section of this chapter. For now it will suffice to conclude that overall welfare reasoning, covering effect of policies in all domains influenced, is part of Planning & Control climate policy in principle. Instrument design should reckon with all such welfare effects. How far this may go in practice is constrained by the practicalities of the complex empirical and political processes involved. In Institutionalism this subject is mostly avoided by creating incentives and creating options, to be filled in at lower level decision levels reckoning with all other circumstances there, including options and constraints created by all other policies.

Energy efficiency policy

One main subject of policy integration is that of energy efficiency policy, serving both the climate policy goal of CO_2 emission reduction and the energy policy goal of security of energy supply. Combining climate and energy policy through efficiency increase has a huge potential according to Grubb, Hourcade and Neuhoff (2014) p. 160), and the same has been said by Hood (2011), as we saw in Chapter 2 concerning "unused energy efficiency options." If policy instruments can be used to speed up specific energy efficiency improvements, rebounding mechanisms will also unavoidably be involved. This is the Jevons paradox. Huge increases in the efficiency of coal production led to an explosion in coal use. Micro-level efficiency improvements may yield less or even have an inverse effect at the macro level, with increased efficiency in energy production and use leading to greater energy use, as occurred for two centuries, from 1750 through 1950. Cost-reducing efficiency improvements also support the huge potential to *increase* energy intensity through various mechanisms (Saunders, 1992, 2000, 2015). Sorrell provides a nuanced analysis, also starting with a review of the Jevons paradox (Sorrell, 2009). More efficient car engines, which have been forced into the market by stricter fleet emissions standards, may not reduce total emissions because they also induce more gasoline use. The average car becomes relatively cheaper. The income effect will first stimulate the sale of more cars and driving longer distances. The efficiency increase in regulated cars will also induce a shift to higher emitting small trucks, now the dominant car in the United States. It will also reduce the share of electric drives, as their cost of driving is not reduced through more efficient combustion engines. The remaining income effect will induce greater general energy use, though, as through more aviation. Aviation is a major domain where people tend to spend additional income (Goodwin, 2012). Even at a micro level such rebound mechanisms may be overwhelmingly important, with sobering outcomes, also in transport (Font Vivanco, Kemp, & van der Voet, 2015; Font Vivanco & van der Voet, 2014). Controlling such rebound mechanisms may be more complex than the original measure that caused the prime intended effect. Climate-relevant rebound mechanisms may be reduced by emission pricing, which can be used to steer activities into less-emitting directions (van den Bergh, 2015). But why use such pricing for rebound mechanisms only? Before exploring these complexities, for reasons of simplicity, options for keeping domains separate should first be considered, also in Planning & Control. Some climate policy instruments may help induce energy efficiency by raising energy prices. Depending on circumstances this might, or might not, lead to increased energy supply security. Focusing on micro-level energy efficiency increases in order to create macro-level energy efficiency serving two goals seems a near-certain route to not serving either goal adequately.

Simplicity in instruments and instrumentation design

Avoiding unnecessary costs in instrumentation

In any welfare-oriented political philosophy, policy instrumentation should be efficient, getting to goals and targets with minimal burdens. These burdens involve not just the private costs of low-emission technologies, but also the private and public regulatory costs of administration and control, and the requirements on the political process, the last two crowding out other policies and other decisions. Constraints on innovation constitute a difficult to quantify effect but long term may be the highest cost component of effective policy. Even incentives-oriented instruments such as cap-and-trade may create barriers to new entrants (Geels et al., 2016; Greiner & Bondarev, 2017), the more so if only partially auctioned for political reasons. Simplicity in instrumentation design covers such aspects: using as few instruments as possible ensures sparseness, while covering all emissions ensures completeness, avoiding application of multiple instruments ensures non-overlap, and not differentiating instruments for the same emissions ensures equality. Moving instrumentation from single installation permits with many instruments to larger groupings, as for fleet standards, reduces the number of applied instruments substantially, exemplifying sparseness. Moving to financial instruments of type 5 will often allow for still broader application, inducing cost-saving innovation as well. Incentive systems may be designed to cover major parts of emissions, such as a cap-and-trade system moving from single-firm emitters to upstream administration at refineries or primary fossil fuel producers, while also covering most transport in the process. Similar lines of reasoning may support design choices for simplicity.

Basic simplicity criteria: sparseness, completeness, non-overlap, equality

Sparseness, as in using fewer instruments, is simpler given that each additional instrument adds complexity in politics, administration and for regulatees. Each addition inevitably creates boundary issues that have to be resolved administrative-technically and politically, requiring additional administrative capacity, which is always limited. This criterion also applies internally within an instrument, such as when differentiations are added to deal with specific characteristics. An international example is in the WTO Environmental Goods Agreement (EGA) which was under negotiation in the Doha Round. It intends to lower tariffs for environmental products, leading to a list of products to be treated equally in terms of reduced tariffs. One of the products listed is air conditioners driven by solar cells. Such a specification requires additional detail: What amount of solar cell is required? Must they be differentiated as to climate zone? May they run on grid electricity when the sun is not shining? Are tariff exemptions also valid if the final buyer and user prove to have solar cells on their roof? Or if they purchase exclusively green electricity? Each question must be answered. But the overall reasoning to answer the why's becomes complex. National authorities are to enact and implement these rules in climate-oriented tax regulations, ultimately for each product on the list, with demarcations and adaptation in other regulations. Sparseness goes against such product differentiations relative to other imported products. As this tax instrument will hardly have a strong influence, other instruments will have to be added for deep emission reduction, not only violating the sparseness criterion but also non-overlap and equality.

Completeness, as in instruments covering all emitters, is the next criterion and works in the opposite direction: emissions not yet covered must be covered by additional instruments, or the domain of application of given instruments must be widened. If a cap-and-trade system is not applied to smaller firms, these smaller firms should be regulated by additional instruments, such as performance standards. This goes against the sparseness and equality principles. Revising a cap-and-trade system to cover smaller firms as well is the other option, such as by shifting the administrative application upstream. The smaller firms are then covered indirectly through the higher price they pay for fossil energy resources.

Non-overlap, as in not covering emitters by more than one instrument, supplements the other criteria. Adding instruments to cover all emissions satisfies the completeness criterion, but violates the sparseness criterion. Avoiding overlap in instrumentation design thus may help reduce the total number of instruments while at the same time still covering all emissions. An example of overlap is renewable electricity support for both wind and solar energy. Wind, solar and nuclear may function in a system with emission pricing — emission tax or cap-and-trade —that makes them more competitive relative to fossil fuels. But at the same time they may also receive subsidies to make them more competitive. Smart solutions can reduce the overlap effect somewhat. The UK subsidy for new nuclear energy (for the new Hinckley Point nuclear power station being built) varies the amount of subsidy depending on the market price of electricity. The sum total of market price plus subsidy is guaranteed. If emission pricing would raise the price of electricity, the subsidy is automatically reduced. If subsidies for wind and solar would reduce electricity prices, the nuclear subsidy goes up. But this price guarantee also reduces the incentive to adapt production to demand, an option that can technically also be developed with nuclear energy (OECD, 2011), certainly if combined with heat production or hydrogen production. Without that incentive renewables face longer periods of very low prices and require greater subsidies to increase their share. The subsidy part of nuclear energy then goes up as well. Overlap in instrumentation can also lead to inequality of treatment, just as incompleteness does in the other direction.

Equality is the most general criterion: all emissions should be treated equally. Each CO_2 molecule emitted plays an equal part in global climate forcing. It does not make a difference if it is emitted from coal, oil, natural gas or peat, nor where the fossil fuels are mined or processed or emitted. From a climate point of view they are equal. If 50% of all emitting sources are closed fully, leaving the other 50% undisturbed at 100% emission, from a climate point of view this is equal to the reduction by 50% of each source. How to define equality?

A basic cost criterion links efficiency to equality. If two sources can each reduce emissions up to 100 tonne, one at the cost of $100 per tonne and the other of $40 per tonne, the overall costs of reducing emissions can vary substantially. The lowest cost of reducing 100 tonne will be $4,000. Equal absolute reduction for both sources, 50 tonne each, will cost $7,000 (2,000 + 5,000). An emission price of $45 per tonne would incentivize only the low-cost emission reduction, 100 tonne, but not the high cost one. The OECD advises that instrumentation be set up as if an ideal tax in ideal markets could be implemented, and then force the outcome on society with any effective instrument (Hood, 2011). Reality is unruly for this task, however, as the private costs of all alternatives have to be known. Dynamics, different for different sources and difficult to predict, make this specification task even more cumbersome and require high flexibility in instrumentation. Applying the equality criterion first helps in filling in the design criteria of sparseness, completeness and non-overlap, and next helps to avoid unnecessary costs in detailing instrumentation.

The four criteria also give some guidance with respect to designing individual instruments. Their main role, however, is in the design of the instrument mix as a whole. How do instruments in a mix relate, such as fleet standards in some transport systems, cap-and-trade in electricity production and in some industries, and energy standards in heating and cooling in buildings? Can the mix be improved by shifting boundaries, or by avoiding overlap and non-coverage? Why develop fleet standards for personal cars but not for trucks, shipping, rail transport and aviation? Or conversely, why not cover them with emission pricing as is done in the energy sector, parts of the industry sector and in aviation? And why cover natural gas use in emission pricing in electricity production, as the EU does, but not most heating in industry and households, and not on natural gas, which increasingly is used in transport? Can instruments be aligned in their cost of emission reduction, including the cost of the dynamics involved, or the lack of these? Or can they be transformed to make them more similar? These four simple, quite technical, design criteria are helpful for posing such questions, and for starting to answer them.

Favoring direct instruments over proxy instruments

In the 1970s, instrumentation on environmental pollution began regulating technologies in operating permits. A main shift was from these operating permits with technical specifications to effect-oriented instruments regulating emissions as in emission standards, and next pricing emission, where emissions could be measured continuously (Huppes, 1988; Huppes & Kagan, 1989). The effect orientation is a prime characteristic relating directly to effectiveness, also a requirement for efficiency measurement. All other instruments are proxy instruments intended to lead to emission reduction through one or more indirect mechanisms. A generic emission price covers combustion engine emissions while a tax on car mass has some relation to emissions and a subsidy on electric cars has a very indirect one, also reckoning with effects upstream in the life cycle as regarding fossil emissions in electricity production. An emission standard in power production works directly while a subsidy on solar PV production through feed-in tariff works through complex indirect mechanisms, highly situation dependent. Under a strict and well-functioning cap-and-trade system, as intended in the EU, the emission reduction from feed-in tariffs is zero. The fragmentation of electricity markets they contribute to may prevent *other* emission reductions, having a net emission *in*creasing effect. But also without such a cap, feed-in tariffs have complex dynamic mechanisms leading only to emission reduction, by first lowering prices of fossil electricity and ultimately reducing prices and hence volumes of primary production of fossil fuels. Such ultimate, longer term, mostly global effects are extremely difficult to quantify, making the specification of effectiveness, and more so efficiency, a near impossibility for such proxy instruments.

In many situations, however, direct instruments are not applicable, barring a generic emission price on CO_2 or an encompassing cap on fossil fuel production in terms of carbon. The option then is to have instruments as close to CO_2 emissions as possible. Fleet standards are linked to emissions under lab standards, with substantial differences as compared with real driving performance per car type involved. Next, they are not linked to driving distance per car and to fleet volume. They are closer to emissions, however, than energy taxes also covering non-fossil energy. The simplicity objective links to this scale from direct instruments to most indirect proxy ones. One instrument for each source of emissions avoids overlap and is complete in coverage. Making the instruments more encompassing reduces their number, increasing sparseness. Having the same costs on all emission reduction creates equality. For proxy instruments such reasoning is less straightforward. Reducing emissions of electricity production by improved building insulation and energy standards for appliances requires many instruments, with limited effects on emissions, certainly when the share of fossil fuels in electricity production is low, as in some Nordic countries now already, and in other countries later assuming effective climate policy.

Formal rules in instrument and instrumentation design

More abstract reasoning can be useful to guide the design process, especially when involving technology-differentiated instruments. Formal reasoning indicates that the number of instruments should be equal to the number of goals to be reached, a rule first formulated by Tinbergen (1952/1970). More instruments will lead to over-determination of the model used to predict effects, without specific outcomes. Fewer instruments will generally make it impossible to reach all goals. This is the Tinbergen rule. Over-instrumentation can be avoided through reducing the number of applicable instruments. Reducing the number of goals, next, can reduce the number of instruments further. One fleet standard for person cars and one for vans and small trucks sets two goals, with two instruments. They may be combined into a single fleet standard, one instrument with one goal, as in the EU, also avoiding arbitrary boundaries at least partly. A seeming solution to instrument number reduction is to cover two goals in one instrument, for example by giving a premium on allowable fleet emissions for also selling full electric cars, as is usual in most fleet standards. The subsidy on electric cars then is hidden in the pricing policy of car companies. This approach not only increase the price of gas guzzling cars and reduces the price of small combustion cars, it also reduces the price of full electric vehicles, to the disadvantage of firms producing full electric vehicles only. These then tend to merge with gas guzzlers, to have that implicit subsidy advantage as well. Such effects upon economic structure favoring incumbent fossil fuel producers are highly undesirable, as well, for reasons of political and administrative feasibility. The two goals can be better served by two instruments, one for reducing emissions and the other for bringing full electric cars into the market. Reducing the number of goals next may be an option. For climate policy it suffices to reduce emissions. How demand reacts to that emission reduction may be left to the private parties involved. They may choose to buy smaller cars, live closer to work, shift to other modes of transport, or buy an electric car. Why engage climate policy with such choice options for emission reduction?

By detailing both goals and targets and levels of government involved, one may then increase the number of instruments, still following the Tinbergen rule (Rio & Howlett, 2013). The consistency analysis then becomes more complex, however, and the policy coordination required, as well. Reconciling international cap-and-trade schemes with national and subnational targets for emission reduction and energy security, for example, is a task that has not yet been performed, and hardly seems doable. The wish to do so very much relates to political philosophy. Planning & Control goes for policy integration covering several goals and administrative levels combined, for integrated welfare reasons. Institutionalism goes for climate effects only, leaving optimization to private parties. Only within a given strategy can goals and targets and instruments be defined, but then very differently between these different political philosophies and their governance strategies. Summarizing, for both strategies it seems wise to keep the number of instruments as low as possible, while covering all sources and avoiding overlap as much as possible. The qualitative guidance is to go for simplicity, defined as *Sparseness, Completeness, Non-Overlap and Equality*.

Coherency in instrumentation design

Operational design: from political philosophy, to simplicity, to resilient effectiveness

Instrumentation design should start from clear political philosophy views, then be as simple as possible, using direct instruments in preference to indirect instruments, third, be capable of reducing emissions effectively and resiliently. Combined, these criteria create coherency in instrumentation. Different political philosophies lead to different instrumentations, as shown in Figure 3.3. But all next-level considerations are substantially open, not dictating one set of instruments as best, and open to differing perspectives of how society may develop

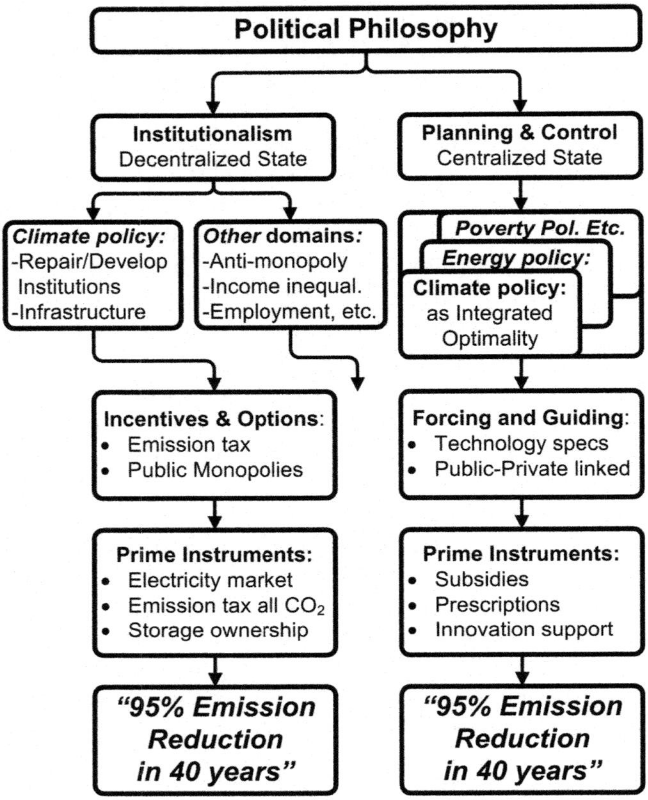

Figure 3.3 From political philosophy to emission reduction.

politically, economically, technically and socially, both nationally and internationally. Design requires imagination and vision, to be daring for long-term relevance.

The first point considers the direction in political philosophy to be followed in the coming decades. Should society develop toward detailed optimized planning and required centralization or toward centralized institutions with decentralized decision making? This is the crossroads. This choice refers not only to climate policy but also to broader constitutional and institutional issues. The clash of political philosophies and the resulting clash of instruments require a clear view of the two main versions present in welfare-oriented political philosophy: Planning & Control and Institutionalism (see Figure 3.3). Including communism and order and stability-oriented conservatism in the liberal-democratic category, the current political spectrum is nearly fully covered. Other political philosophies with perspectives on effective climate policy are not yet in view. Might they arise? A stable authoritarian society is not necessarily linked to collective welfare but may try to avoid the instabilities of climate change. Similarly, decentralized anarchist solutions, as in the "degrowth" movement, accept a substantial reduction in market-based welfare, with concomitant emission reduction, even if not focused at emission reduction.

The second point involves simplicity in instrumentation. This goal relates to effectiveness and efficiency at a more conceptual level. Current climate policy instruments tend to overlap and oppose each other, clogging the regulatory system with overlapping and counteracting instruments in trying to be effective. Planning & Control may have this tendency. But it should be strongly opposed for reasons of effectiveness and efficiency, guided by simplicity. Institutionalism has less of this tendency but must adhere to the same four simplicity principles as well.

The third point encompasses instrumentation design, according to basic political philosophy and simplicity. It has two elements: the starting points for climate policy instrumentation and the role of other policy domains in climate policy instrumentation (see next section on policy integration). Planning & Control takes the main emitting technologies as its starting point, grouped into a few main sectors. Technology-specific instruments aim to reduce emissions from these technologies and help create the low-to-zero emission technologies to replace them. The starting point is at the bottom of Figure 3.4, moving upward only where necessary for technology-specific issues, mostly toward RD&D, with emphasis on Demonstration. All instruments integrate with other goals in society, primarily safe energy supply and poverty. Institutionalism looks at deficiencies that arise in the regulatory framework of society. It repairs institutions that have become inadequate and creates new ones where required. There is no sectoral analysis. Changes are induced by generic incentives and option creation. The starting point is at the top of Figure 3.4, going downward only if needed, not beyond RD&D, type 3 and with emphasis on Research there. What exactly the outcomes will be for specific technologies is left to lower level public bodies, civil society, firms, private persons, and markets as they develop, all influenced

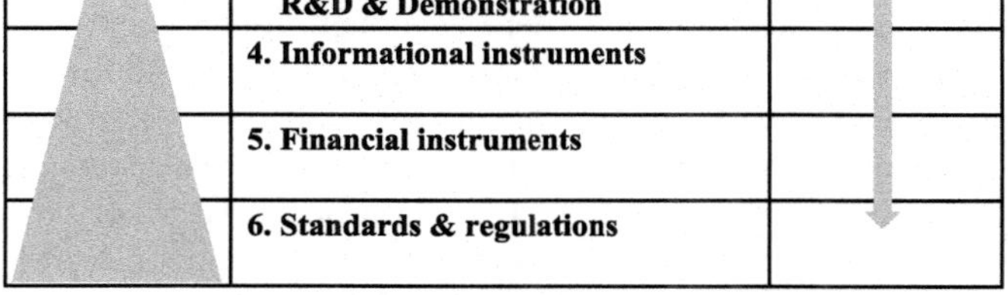

Figure 3.4 Specific and generic instrumentation design strategies.

by the revised institutions. The innovations in institutions are focused here on climate change, only, embedded in the broader framework of society. Other domains of change will require their own novel institutional developments, such as in reducing income inequality and tendencies toward economic and political monopolization. Their conjoint effects may be both supportive and contradictory, depending on domains and subjects, and more detailed choices made within them. Only at a most abstract level may there be normative integration, guiding institutional repair to the same philosophical background, as in safeguarding individual independence in an integrated but open society.

Resilience in instrumentation design has a number of contradictory elements. It is first based on avoiding premature choices and assumptions regarding technologies; not betting on specific assumptions as regarding negative emissions and avoiding lock-ins to emitting technologies by delaying choices. Next it exerts pressure for fast enough emission reduction to remain within the 2/1.5 degree realm of temperature rise, forcing the relevant technologies to exit and force in the new zero-emission ones on time. A core resilience issue is political and administrative in nature. Can the relevant instrumentation be decided upon and implemented, and can effective instrumentation be maintained in the next decades, up to half a century at least, when other issues also climb the political agenda? The triad in design choices may help: from political philosophy to structured design, always going for simplicity.

Policy integration and co-benefits differ between political philosophies

There are deep interrelations of values and goals, as in the Sustainable Development Goals (SDGs), and there are interactions in reaching them. Ideally, all have their place in all decision making, together contributing to maximizing individual and collective welfare in their broadest sense. Intellectually the framework is there, as exemplified in in the work of Stiglitz, Hourcade, Nussbaum and Sen, introduced earlier in Chapter 2. However, how to define this laudable welfare goal is hardly possible in a practical sense. How to reach it is not only difficult for that reason. The means, the instruments, cannot approach the ideal directly due to the undetermined nature of human desires. Nobody has a well-defined current and future utility function, let alone one that can be integrated into a dynamically defined social welfare function operationally. The two political philosophies supply different solutions to this fundamental policy problem. Planning & Control assumes there is enough knowledge on utility functions to move society in the right direction, though not based on the dynamic welfare function of society but on partial reasoning, as a type of cost-benefit analysis assuming, 'all other things to be equal'. Specific decisions can then be clear, within a limited time horizon, short term optimization of Grubb (2014). Subsidies and taxes to introduce renewables hurt lower income groups disproportionally, certainly those in substandard dwellings. They are to be compensated financially and through insulation programs of buildings, which increase the safety of the energy supply, as well. Such policy integration looks at options in which combined improvements can be realized, where co-benefits can be created, and co-cost can be avoided or compensated. Examples of co-benefits are effects on health (Jack & Kinney, 2010) and employment (Fankhaeser, Sehlleier, & Stern, 2008). Co-costs are effects on income distribution and poverty, for example (Adler et al., 2017). Full policy integration is replaced by practical solutions, looking for combined but partial improvements. If other domains follow the same approach, climate policy is mainstreamed there, symmetrical to other policy domains being mainstreamed in climate policy instrumentation. In the somewhat softer co-benefit/co-cost line of reasoning, a practical cutoff has to be defined, beyond which other policy domains have priority in taking account of climate issues, if deemed relevant and attractive from a combined but still partial welfare point of view. Such integration, as mainstreaming seen from a climate policy point of view, is advocated in Planning & Control (Kok & De Coninck, 2007; Rauland & Newman, 2015; United Nations Development Programme [UNDP], 2012).

In Institutionalism, repairs and redesign of institutions in the climate domain are due. But they should leave the overall superstructure largely intact: the constitution, institutions and main infrastructure. Changing all or even a substantial part is simply not possible. Fitting in with this overall societal fabric allows other issues in other domains to be resolved in a similar way, as well, by domain-specific institutional adaptations. There is integration, but not by instrumentation at the level of technology-specific actions. In the environmental pollution domain, health effects have partly been brought into ownership rules by adapting liability rules, as in creating joint-and-several liability. However, this legal adaptation has been instrumental in preventing local soil pollution, only. Such legal liability solutions must be investigated to maintain unity in institutions. But this approach is not applicable in the climate domain. With institutional repairs enacted, action integration is still required. That is not the task of central government, only indirectly that of lower level government, but mainly private parties, guided by the decentralized wishes of persons and organizations, including civil society and firms, enabled and guided by the adapted incentive structure (Peng & Bai, 2018; Rauland & Newman, 2015).

Summarizing, co-benefits creation and also broader policy integration are key ingredients in Planning & Control climate policy instrumentation, which focuses on more micro-level decisions. They are not part of Institutionalist instrumentation. Integration there is by adhering to the main institutional structure of society, fitting in with its main principles, leaving actual integration to private parties and lower level governments.

Combining the two instrumentation strategies?

As both strategies for instrumentation have a similar welfare background, individuals count, and as both adhere to simplicity, there might be some common ground in instrumentation. A point of overlap might, for example, relate to the variably priced electricity market. These are useful for moving in emission saving technologies such as electric cars and heat pumps for heating and cooling. Planning & Control would then move somewhat toward Institutionalism, which advocates real-time electricity pricing. However, other instruments in Planning & Control focus on subsidized renewables and capacity payment for reasons of security of supply. Real-time equal for all electricity pricing is hardly possible, then, and loses much of its advantages. Also, subsidizing specific industries for not emitting is a fundamentally wrong approach to take in Institutionalism. It goes against the polluter pays principle and related liability considerations, and requires raising other taxes to fund these inefficient outlays. Some practical overlap may occur between the fundamentally separate strategies.

Strategic instrumentation design for Planning & Control and Institutionalism

Planning & Control starts with better regulation of private actors toward emission reduction. It uses prescriptions as standards and regulations (type 6), subsidies and taxes as financial incentives (type 5), information to improve private choices (type 4), and next involves public R&D and Demonstration (type 3). Infrastructure and spatial planning (type 2) follow only as needed, with, for example, rules on public-private partnerships in both demonstration and infrastructure development. Institutional framework adaptations (type 1) hardly appear on the agenda. The Planning & Control process starts at the bottom, at the left of Figure 3.4.

Institutionalism starts at the most general level: institutions, including market development (type 1). It then adds public infrastructure and spatial plans (type 2), but only where the climate aspect plays a primary role. It supports basic research and some applied research (type 3), mostly from a more general point of view than specific technologies. Only in very exceptional circumstances is there support for creating learning curves for very promising new technologies, and then when their market introduction is blocked by strong lock-ins to existing technologies and where high risks are involved. Technology-specific regulations are an exception, existing only where markets are deeply deficient and normal market repair or development is not possible; see the right column in Figure 3.4.

References

Acemoglu, D., & Robinson, J. A. (2012). *Why nations fail: the origins of power, prosperity and poverty*. New York: Crown Publishers.

Adler, M., Anthoff, D., Bosetti, V., Garner, G., Keller, K., & Treich, N. (2017). Priority for the worse-off and the social cost of carbon. *Nature Climate Change*, *7*(6), 443–449.

Ambec, S., Cohen, M. A., Elgie, S., & Lanoie, P. (2013). The Porter Hypothesis at 20: Can environmental regulation enhance innovation and competitiveness? *Review of Environmental Economics and Policy*, *7*(1), 2–22. doi: 10.1093/reep/res016

Baumol, W. J. (1972). On taxation and control of externalities. *American Economic Review*, *62*(3), 15.

Baumol, W. J., & Oates, W. E. (1971). The use of standards and prices for protection of the environment. *Swedish Journal of Economics*, *73*(1), 12.

Boldrin, M., & Levine, D. K. (2008). *Against intellectual monopoly*. Cambridge, England: Cambridge University Press.

Brown, R., Gregson, G., & Mason, C. (2016). A post-mortem of regional innovation policy failure: Scotland's Intermediate Technology Initiative (ITI). *Regional Studies*, *50*(7), 1260–1272. doi: 10.1080/00343404.2014.985644

Calabresi, G., & Melamed, A. D. (1972). Property rules, liability rules, and inalienability: one view of the cathedral. *Harvard Law Review*, *85*(6), 1089–1128.

Cyert, R. M., & March, J. G. (1963). *A behavioral theory of the firm*. Englewood Cliffs, NJ: Prentice-Hall.

European Commission. (2009). *Directive 2009/72/EC of the European Parliament and of the Council of 13 July 2009 concerning common rules for the internal market in electricity and repealing Directive 2003/54/EC (Text with EEA relevance)*. Retrieved from http://data.europa.eu/eli/dir/2009/72/oj.

European Commission. (2013). *Ownership unbundling. The Commission's practice in assessing the presence of a conflict of interest including in case of financial investors*. Brussels: European Commission.

Fankhaeser, S., Sehlleier, F., & Stern, N. (2008). Climate change, innovation and jobs. *Climate Policy*, 8(4), 421–429.

Faure, M., & Peeters, M. (2011). *Climate change liability*: Northampton, MA: Edward Elgar Publishing.

Font Vivanco, D., Kemp, R., & van der Voet, E. (2015). The relativity of eco-innovation: environmental rebound effects from past transport innovations in Europe. *Journal of Cleaner Production*, *101*, 71–85. doi: 10.1016/j.jclepro.2015.04.019

Font Vivanco, D., & van der Voet, E. (2014). The rebound effect through industrial ecology's eyes: a review of LCA-based studies. *The International Journal of Life Cycle Assessment*, *19*(12), 1933–1947.

Foxon, T. J. (2011). A coevolutionary framework for analysing a transition to a sustainable low carbon economy. *Ecological Economics*, *70*(12), 2258–2267. doi: 10.1016/j.ecolecon.2011.07.014

Geels, F. W. (2002). Technological transitions as evolutionary reconfiguration processes: a multi-level perspective and a case-study. *Research Policy*, *31*(8–9), 1257–1274. doi: 10.1016/S0048-7333(02)00062-8

Geels, F. W. (2014). Regime resistance against low-carbon transitions: introducing politics and power into the multi-level perspective. *Theory, Culture & Society*, *31*(5), 21–40. doi: 10.1177/0263276414531627

Geels, F. W., Kern, F., Fuchs, G., Hinderer, N., Kungl, G., Mylan, J., … Wassermann, S. (2016). The enactment of socio-technical transition pathways: a reformulated typology and a comparative multi-level analysis of the German and UK low-carbon electricity transitions (1990–2014). *Research Policy*, *45*(4), 896–913. doi: 10.1016/j.respol.2016.01.015

Geels, F. W., & Schot, J. (2007). Typology of sociotechnical transition pathways. *Research Policy*, *36*(3), 399–417. doi: 10.1016/j.respol.2007.01.003

Geels, F. W., Sovacool, B. K., Schwanen, T., & Sorrell, S. (2017). Sociotechnical transitions for deep decarbonization. *Science*, *357*(6357), 1242–1244. doi: 10.1126/science.aao3760

Goodwin, P. (2012). *Peak Travel, Peak Car and the Future of Mobility. Evidence, Unresolved Issues, Policy Implications, and a Research Agenda.* Paper presented at the Roundtable on Long-Run Trends in Travel Demand, 29–30 November 2012.

Greiner, A., & Bondarev, A. (2017). Optimal R&D investment with learning-by-doing: multiple steady states and thresholds. *Optimal Control Applications and Methods, 38*(6), 956–962.

Grubb, M., Hourcade, J. C., & Neuhoff, K. (2014). *Planetary economics: energy, climate change and the three domains of sustainable development.* Abingdon, UK: Taylor and Francis.

Hamamoto, M. (2011). Energy efficiency regulation and R&D activity: a study of the top runner program in Japan. *Low Carbon Economy, 2*(02), 91.

Hirooka, M. (2006). *Innovation dynamism and economic growth: a nonlinear perspective:* Northampton, MA: Edward Elgar Publishing.

Hogan, W. W. (2014). *Time-of-use rates and real-time prices.* Retrieved from https://sites.hks.harvard.edu/fs/whogan/Hogan_TOU_RTP_Newark_082314.pdf.

Hood, C. (2011). *Summing up the parts. Combining policy instruments for least-cost climate mitigation strategies.* Paris: OECD/IEA.

Huppes, G. (1988). New instruments for environmental policy: a perspective. *International Journal of Social Economics, 15*(3/4), 42–50. doi: 10.1108/eb014102

Huppes, G., & Kagan, R. A. (1989). Market-oriented regulation of environmental problems in the Netherlands. *Law & Policy, 11*(2), 215–239. doi: 10.1111/j.1467-9930.1989.tb00027.x

Huppes, G., & Simonis, U. E. (2009). Environmental policy instruments. In J. J. Boersema & L. Reijnders (Eds.), *Principles of environmental sciences* (pp. 239–280). Dordrecht, Netherlands: Springer.

Jack, D. W., & Kinney, P. L. (2010). Health co-benefits of climate mitigation in urban areas. *Current Opinion in Environmental Sustainability, 2*(3), 172–177.

Jaffe, A. B., & Lerner, J. (2011). *Innovation and its discontents: How our broken patent system is endangering innovation and progress, and what to do about it.* Princeton, NJ: Princeton University Press. (Originally published 2004.)

Joore, P., & Brezet, H. (2015). A multilevel design model: the mutual relationship between product-service system development and societal change processes. *Journal of Cleaner Production, 97*, 92–105. doi: 10.1016/j.jclepro.2014.06.043

Kok, M., & De Coninck, H. (2007). Widening the scope of policies to address climate change: directions for mainstreaming. *Environmental Science & Policy, 10*(7), 587–599.

Kramer, G. J., & Haigh, M. (2009). No quick switch to low-carbon energy. *Nature, 462*(7273), 568–569.

Lindblom, C. E. (1959). The science of 'muddling through'. *Public Administration Review*, 79–88.

Mancinelli, S., & Mazzanti, M. (2009). Innovation, networking and complementarity: evidence on SME performances for a local economic system in north-eastern Italy. *The Annals of Regional Science, 43*(3), 567–597.

Mazzucato, M. (2013). *The entrepreneurial state: Debunking public vs private sector myths.* London: Anthem Press.

Mingardi, A. (2015). A critique of Mazzucato's entrepreneurial state. *Cato Journal, 35*, 603.

Nishitani, K., & Itoh, M. (2016). Product innovation in response to environmental standards and competitive advantage: a hedonic analysis of refrigerators in the Japanese retail market. *Journal of Cleaner Production, 113*, 873-883

OECD. (1972). *Recommendation to the council on guiding principles concerning international economic aspects of environmental policies*. Paris: OECD.

OECD. (2011). *Technical and economic aspects of load following with nuclear power plants*. Paris: OECD.

Peng, Y., & Bai, X. (2018). Experimenting towards a low-carbon city: policy evolution and nested structure of innovation. *Journal of Cleaner Production, 174*, 201–212.

Pigou, A. C. (1932). *The economics of welfare* (downloadable at: http://www.econlib.org/library/NPDBooks/Pigou/pgEW.html). Indianapolis: Library of Economics and Liberty. (Originally published 1920.)

Porter, M. E., & Van der Linde, C. (1995). Green and competitive: ending the stalemate. *Harvard Business Review, 73*(5), 120–134.

Rauland, V., & Newman, P. (2015). *Decarbonising cities: mainstreaming low carbon urban development*. Cham, Switzerland: Springer International.

Rio, P. D., & Howlett, M. (2013). Beyond the "Tinbergen rule" in policy design: matching tools and goals in policy portfolios. *Annual Review of Policy Design, 1*.

Rockström, J., Gaffney, O., Rogelj, J., Meinshausen, M., Nakicenovic, N., & Schellnhuber, H. J. (2017). A roadmap for rapid decarbonization. Emissions inevitably approach zero with a "carbon law". *Science, 355*(6331), 1269–1271.

Roozenburg, N. F. M., & Eekels, J. (1995). *Product design: fundamentals and methods* (2nd ed. 1998). Chichester, England: Wiley.

Saunders, H. D. (1992). The Khazzoom-Brookes postulate and neoclassical growth. *The Energy Journal, 13*(4), 131–148.

Saunders, H. D. (2000). A view from the macro side: rebound, backfire, and Khazzoom–Brookes. *Energy Policy, 28*(6–7), 439–449. doi: 10.1016/S0301-4215(00)00024-0

Saunders, H. D. (2015). Recent evidence for large rebound: elucidating the drivers and their implications for climate change models. *The Energy Journal, 36*(1), 23–48.

Simon, H. A. (1956). Rational choice and the structure of the environment. *Psychological Review, 63*(2), 129.

Smith, A., Voß, J.-P., & Grin, J. (2010). Innovation studies and sustainability transitions: the allure of the multi-level perspective and its challenges. *Research Policy, 39*(4), 435–448. doi: 10.1016/j.respol.2010.01.023

Sorrell, S. (2009). Jevons' paradox revisited: the evidence for backfire from improved energy efficiency. *Energy Policy, 37*(4), 1456–1469. doi: 10.1016/j.enpol.2008.12.003

Stiglitz, J. E. (2014). *Intellectual property rights, the pool of knowledge, and innovation*. (Working Paper No. 20014). Cambridge, MA: U.S. National Bureau of Economic Research.

Tinbergen, J. (1970 (1952)). *On the theory of economic policy*. Amsterdam: Elsevier Science & Technology.

UNDP. (2012). *Mainstreaming climate change in national development processes and UN country programming: a guide to assist UN country teams in integrating climate change risks and opportunities*. New York: UNDP.

van den Bergh, J. C. J. M. (2015). Climate summit: pricing would limit carbon rebound. *Nature, 526*(7572), 195–195. doi: 10.1038/526195a

Wang, Q., Zhang, C., Ding, Y., Xydis, G., Wang, J., & Østergaard, J. (2015). Review of real-time electricity markets for integrating distributed energy resources and demand response. *Applied Energy, 138*, 695–706. doi: 10.1016/j.apenergy.2014.10.048

Weitzman, M. L. (2009). On modeling and interpreting the economics of catastrophic climate change. *Review of Economics and Statistics, 91*(1), 1–19.

Weitzman, M. L. (2011). Fat-tailed uncertainty in the economics of catastrophic climate change. *Review of Environmental Economics and Policy*, 5(2), 275–292. doi: 10.1093/reep/rer006

Yi, D., Pineda, S., Nyeng, P., Ostergaard, J., Larsen, E. M., & Qiuwei, W. (2013). Real-time market concept architecture for EcoGrid EU-2014; a prototype for European smart grids. *Smart Grid, IEEE Transactions on*, 4(4), 2006–2016. doi: 10.1109/TSG.2013.2258048

4 Designing Planning & Control climate policy instrumentation

Design philosophy: controlled emission reductions

The basic philosophy of Planning & Control is to steer markets and technologies in a planned and controlled way toward near-zero CO_2 emissions. This climate goal is integrated with other policy goals when adapting and linking instrumentation, especially with respect to energy supply, income distribution and poverty alleviation, both nationally and internationally. The measuring rod is an increase in general welfare, broadly defined to include considerations of justice and personal development (Stiglitz, Sen, & Fitoussi, 2009). The 17 UN Sustainable Development Goals (SDGs) have a similar integrated background. Goals and specific targets are set for society, sectors and technologies; next, instruments are specified and implemented to reach the targets, primarily using instruments of types 4, 5 and 6; see Table 3.3 earlier. Where technology-specific action is not possible or is insufficient, a less direct approach to emissions reduction comes into view, especially regarding type 3, publicly (co-)funded development and demonstration and market introduction.

Instrumentation design must take practical limitations into account. Volkswagen's "Dieselgate" is an example in which the regulatory capacities of government were overestimated. The controls on technologies were not effective and were open to manipulation and vulnerable to political influence by the private firms involved. Such practical incapacity can be substantially avoided through application of the ***simplicity*** principle. The regulatory capacity of a society defines the limit of what its political and administrative systems can handle and also defines what the private sector will be willing to accept. This is so even if the 2-degree climate goal was agreed upon earlier. Sunstein's *Simpler: The Future of Government* (2014), indicates how complexity in regulations can be reduced, such as by adding nudging in instrumentation. However, because new and revised regulations relate to other existing regulations, overall regulation tends to become ever more complex and detailed, similar to improvements in large software systems. Simplification is an important but Sisyphean task.

Dealing with major trends

Alignment with major societal developments is the starting point for instrumentation design. In the language of design theory, societal developments constitute the regimes and landscapes of instrumentation design. Climate policy has such a pervasive influence on society that it does not merely adapt to societal developments, but is a primary constituting factor of social developments. The major trends for the next decades that are relevant and significant from a Planning & Control perspective include the following.

- Globalization of information, technologies and most markets will continue.
- Patent protection will be increasingly international.
- Learning by doing will be key to market development.
- Public–private partnerships will be key to development of new technologies and infrastructure.
- Income inequality will increase further, with a role for climate policy in its reduction.
- Decreasing legitimacy will be reversed by direct voting procedures and internet participation.
- Many small steps will be the key to long-term improvement.
- Global climate policy will set binding emission reduction targets per country.
- Country targets will reflect GDP per head, and also past emissions, for reasons of justice and equity.

Setting targets for reduction

For the world as a whole, the carbon budget for remaining within the 2-degree climate warming goal was around 800 gigatonnes of CO_2 ($GtCO_2$) in 2015, and around 400 $GtCO_2$ for the 1.5-degree target (Höhne et al., 2017; Millar et al., 2017; Rogelj et al., 2016). Emissions since 2015 have been close to 40 $GtCO_2$ per year. At this rate, the budget would be spent in 20 years' time, leaving it at zero thereafter. Stabilization at 2-degrees or preferably 1.5 degrees of temperature rise requires instruments capable of reducing emissions by 95% in 2050 and to zero by 2060, certainly in already industrialized countries. Large scale bioelectricity may have higher greenhouse gas emissions than natural gas or even coal (Stephenson & MacKay, 2014). Carbon capture and storage (CCS) cannot be reliably counted on, nor can direct air capture and storage (DACS). Climate policy instrumentation must therefore reduce all fossil fuel use to near zero, replacing it with non-fossils. Energy needs are not fixed. The United States and European Union have a similar level of economic development but differently priced energy systems, and volumes of CO_2 emissions per unit of GDP that differ by a factor greater than 2 (Grubb et al., 2018). Higher energy pricing reduces real cost by increasing efficiency.

The 95% reduction target set for developed countries by 2050 must be translated into targets for all major sectors. Sectors do not function independently of each other, however. Achieving zero direct emissions in transport, including aviation, would leave only its upstream emissions. All material processes for producing vehicles, roads, airports, and related technology now amount to around one quarter of emissions in the life-cycle chains of transport. Even with zero direct emissions in transport, the overall reduction target of minus 95% cannot be reached without adequate measures being undertaken in other sectors as well. Zero direct emissions are relatively easy to achieve in personal transport as electric drives and hydrogen fuel cell drives can achieve that already. The transition to zero direct emissions in the buildings, industrial and energy sectors is a much more challenging task.

Master plan for instrumentation

Each country must make a master plan for emissions reduction that specifies the emissions per year, remaining in adherence with international agreements and pledges. Emissions may first be reduced by taxing them or by subsidizing their reduction, the two options provided under the Coase theorem (Coase, 1960). Zero-emission technologies that are available and not too expensive can be subsidized or prescribed to replace fossils. Where starting costs are high, focused innovation can also be organized and funded. The stringency required from technology-specific instruments for the ultimate zero-emission goal implies deep interrelations with other technologies and with non-climate goals and objectives. Requirements for a secure energy supply are fundamentally influenced by large-scale introduction of intermittent renewables and the reduction of the fossil fuel supply, requiring compensating actions such as backup power stations, energy storage and spreading of peak demand. Deep adaptations in household heating and cooling will require costly investments, which are most burdensome for lower income groups, along with an early write-off of currently invested capital goods and appliances. The emission reduction task is thus transformed into technology and market change tasks. The reduction tasks will require effective instruments for implementation. A country's decision to prohibit the sale of combustion-drive cars after 2040 is a clear and simple target and part of the climate goal. But the reduction will have to be gradual, along a predefined path, using instruments still to be developed and to be decided on. Instrumentation in Planning & Control develops from the reduction paths set, with economic-technical analysis detailing targets that may differ to some extent from country to country. The long-term challenge is the same for all countries: to reduce the use fossil fuels in most applications to zero.

Simplicity

Instrumentation design next aims for simplicity, with *sparseness, completeness, non-overlap and equality* as operational guidelines, as detailed in Chapter 3. The prime choice is for instruments that regulate emissions directly, the long-standing ideal in environmental policy. Instruments that use proxy variables reduce emissions indirectly, having more complex mechanisms and more side-effects. These instruments should be as flexible and incentivizing as possible, with information as the least binding option, followed by taxes and subsidies pressuring towards emission reduction, then by tradable emission permits with restrictions for groups of emitters, then by more rigid emission standards and lastly by technology prescriptions. An emission standard per kilowatt hour (kWh) of electricity produced is simpler than prescribing a combined cycle for gas-fired power stations, as the standard applies to all fossil power stations.

Encompassing system-level instrumentation

A systems-level certificate instrument for planned emissions reduction

The simplest and most direct regulatory instrument is the limitation of fossil inflow into the economy and expansion of fossil outflow, regulated by a fossil inflow certificate system. Any given country's emissions are equal to the total inflow by primary production plus imports minus total outflow in exports and back to geological storage; see Figure 4.1. Each primary producer and importer must be in possession of inflow certificates. Certificates are provided by auction equal to total emissions for the year. Certificates are earned with exports of fossil energy products and with geological storage of carbon, whether fossil or biotic.

The volume of auctioned certificates is reduced by a predetermined percentage each year, in line with strict planning reasoning as in Rockström and co-authors (2017). For a 90% emissions reduction in 40 years' time, the scheme may start slowly to avoid undue costs, beginning with a 0.5% reduction during the first year,

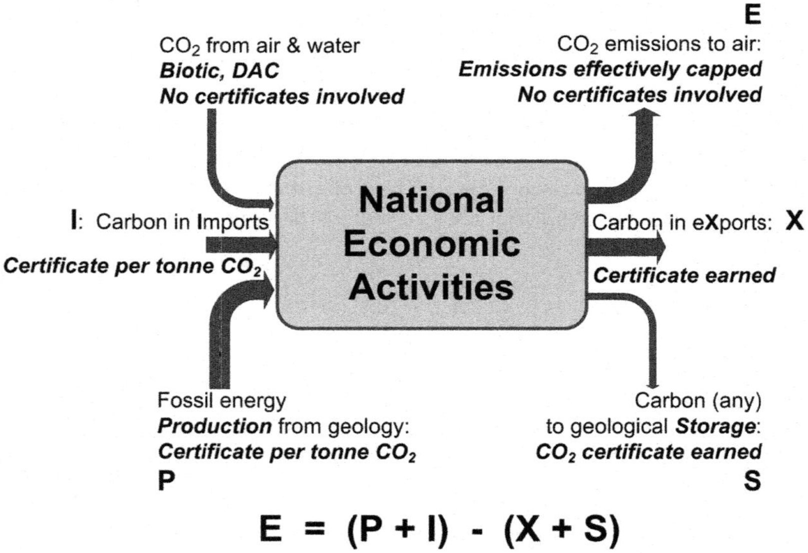

Figure 4.1 Fossils certificates for primary production and imports and CCS and exports.

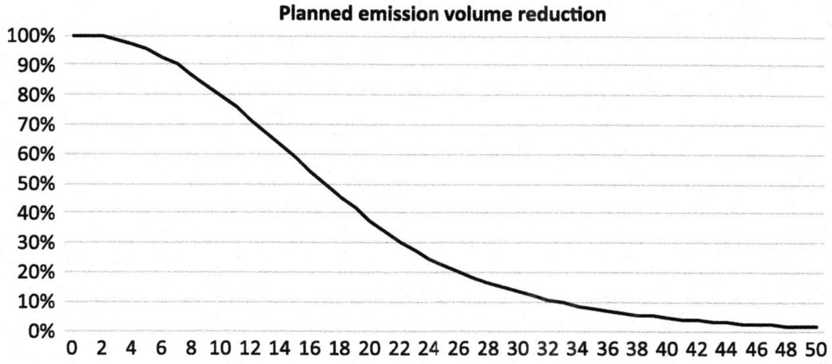

Figure 4.2 Emissions volume reduction by 95% over 40 years.

and rising by an additional 0.5% each year thereafter. Starting at 40 Gt emissions globally, the reduction in year 2 would be 0.8 Gt. In year 15, the yearly reduction rate would reach its maximum of 7% (compound rate), from then onward halving emissions every decade. In absolute terms, the reduction starts slowly, rises to its maximum in year 15, and then slows down as the fixed percentage refers to an ever-decreasing volume; see Figure 4.2 The timeline is not a prediction or a prospective scenario, but rather is a politically decided standard. Certificates not yet used in the period of supply are reduced in volume by the same percentage to avoid impossibly high reductions later.

Importers and primary producers buy certificates in real time from the issuing governmental body and from those storing carbon permanently or exporting it. Some flexibility is created by making the registered certificates tradable on a public market. Administration of certificate use is conducted on a real-time transaction basis, as is now common with duties and excises such as on alcohol and many import products. However, here the volume is fixed, not the price. The total emissions reduction is thus predictable for the purpose of planning. The timeline is set so as to avoid long-term price peaks and troughs. Uncertain economic developments and novel technological developments will create short- and medium-term price changes that are unavoidable and difficult to predict. Banking can play a limited role as each certificate loses some of its value over the course of time. Cumulatively, a certificate issued in year 1 will have depreciated by 35% in year 10. For very sensitive industries, trade may include futures and options to hedge against the risks of large price fluctuations.

The proceeds of bringing certificates to market go to the public provider of certificates, to private exporters and to the providers of geological storage of CO_2. Certificates may be issued by lower administrative levels of government, such as by countries in the EU, provinces in Canada and China and states in the United States, but higher level administration reduces costs. Two or more countries, states or provinces may agree to conduct trade free of certificates. Together they then act as one country. In this scenario, proceeds of certificate sales are distributed among these entities according to the ratio of their actual emissions, creating incentives against underreporting.

Certificates compared with cap-and-trade and encompassing emission tax

At first view the system-level certificate cap seems like a cap-and-trade system comparable to the EU-ETS and similar systems in place in many countries, states and provinces. However, basic differences with cap-and-trade can be distinguished:

- The certificates are required for primary fossils producers and importers, not emitters.
- They cannot be linked to specific emitters, which reduces political vulnerability.
- Emitting firms or persons do not require an emission administration or certificate administration to be controlled by government ex-post, easing administrative requirements significantly.
- Emissions from all sectors are covered, including all forms of transport.
- Smaller emitters are also equally covered through the same market mechanisms under which they already function.

The system-level certificate system may also be compared with a predictably rising upstream-administered emission tax at a systems level, as discussed later in Chapter 6, with three main differences:

- The certificate price is not fixed, making investments more risky than with a predictably rising emission tax.
- Other Planning & Control instruments, such as subsidies for renewables, will lower the certificate price but not emissions, in contrast to the tax.
- Trade-in certificates allows for "tax optimization,", as it is easy to shift profits and losses to where profits are taxed less and losses can be deducted from income before taxes. This is made easy as no physical items are involved, similar to intellectual property rights proceeds used for the same purpose.

The certificate system and the system-level emission tax also share basic similarities:

- Both need to administer only a limited number of firms and not the actual emitters.
- They function on a real-time transaction basis and not at the delayed administrative control of cap-and-trade, usually more than a year after emission.
- The cover all CO_2 emissions, including those from smaller and mobile sources.
- They generate roughly the same proceeds, calculated over the years.

The quantity-oriented systems approach with fossils certificates may well accord with an abstract approach toward planning emissions reduction but is not linked to a specific reduction path. Simplicity reasoning tends toward increased effectiveness and efficiency, such as through stabilizing prices and avoiding auctions and banking, moving in an Institutionalist direction. Planning & Control reasoning, in contrast, tends toward the more detailed democratic welfare orientation, linked to more technology-specific planning. Such planning will link to transforming primary technologies in main sectors for the purpose of realizing the transition toward a zero-emission society, making the certificate system a superfluous burden. The system-level emission cap using certificates goes a step too far in the reasoning of Planning & Control, and falls a step short for Institutionalism.

Guide for design of regulatory instruments per sector

Intersectoral linkages: demand pull and supply push

In the end-use sectors, all energy production, energy use and emissions are demand driven. The energy sector itself supports use in the other three sectors. Changing the types and volumes of final demand – demand pull – is one main entry into instrumentation. If transport systems shift to non-fossil drives, demand reduction will lead to reduced production of transport fuels at refineries, refineries will take in fewer fossil fuels and the primary production of oil, coal and natural gas will decrease. Similarly, if heating in the building sector shifts from natural gas to electricity-based heat pumps, direct demand for natural gas will decrease as well as its production and price, while electricity demand goes up. While rebound and other economic mechanisms may reduce the overall effectiveness of specific shifts in final demand, demand pull is nonetheless a primary entry for climate policy instrumentation and should be applied broadly using direct instruments, or proxy instruments where direct instruments do not suffice. Demand pull cannot influence the upstream technologies used for energy supply, however. Shifting from natural gas to electricity for heating will not influence the way electricity is produced. The expansion of electricity production over the past decades has mainly been through building coal and natural gas– based power stations, with renewables playing a very limited role. Increased energy efficiency (Patterson, 1996) might reduce energy use, or not – the Jevons paradox – but it will have little impact on how upstream production takes place.

Demand pull must be accompanied by supply push: changing primary energy production by crowding out fossils and bringing in non-fossil energy: solar, wind, hydro, geothermal, wave, nuclear, and others. Bringing in non-fossils, such as by subsidies, although widely used, crowds out only part of fossil supply as induced lower prices discourage only more expensive production. Technology-specific instruments per sector must reckon with total system effects caused by combined supply-and-demand mechanisms, as an ideal emission tax would do (Hood, 2011).

Instruments for a zero-emission energy sector, producing zero-emission fuels

The energy sector surveyed

The energy sector mainly transforms fossil fuels into energy carriers such as electricity and fuels for intermediate and final use in the industrial, buildings and transport sectors. See Figure 1.5, earlier, showing the complexity of this sector in the EU, and Figure 4.3, showing in the grey area what is to disappear. A detailed analysis of the energy sector is provided in the fifth assessment report of the Intergovernmental Panel on Climate Change (IPCC-AR5-WGIII, 2014, chapter 7).[1] Virtually all current fossil-based activities in the energy sector will have to be closed down in the coming decades.

The boundaries between the energy sector and other sectors will become ever more fluid due to both more non-energy outputs of the energy sector and more energy inputs to the energy sector from industry, as in biomass and hydrogen production; from transport, as in stored energy in vehicle batteries; and from the buildings sector, with primary and secondary electricity. A narrow energy sector analysis may overlook more encompassing options for emission reduction. If one instrument is applicable to activities across several sectors, applying it in one sector, only, would violate the principles of *sparseness* and *equality*. The IEA advice in this regard is to assess how a generic emission tax would work over all sectors, and to implement that result (Hood, 2011). However, the assumed technical outcomes under the same tax would differ significantly between regions and climate zones, and between countries.

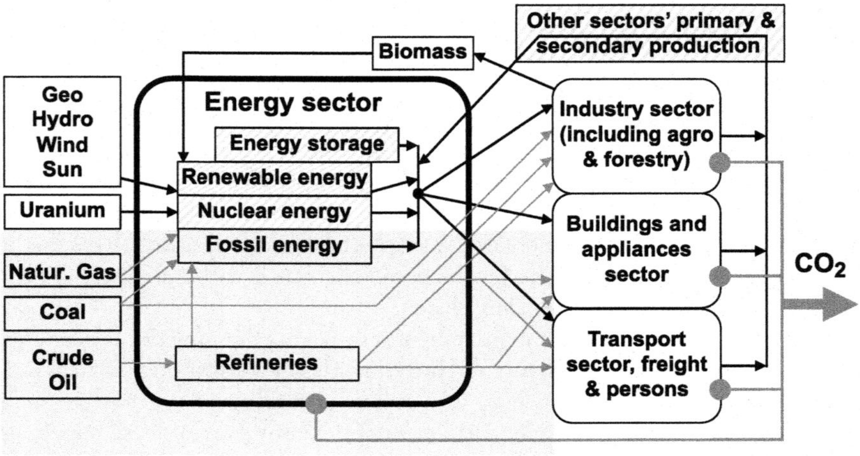

Figure 4.3 Energy sector diagram.

Direct instruments for emission reduction in the energy sector

In order to avoid the costs of disruptive change, instruments will only gradually phase out fossil electricity and heat. The basic options include instruments targeting emissions directly and those working indirectly on proxy variables; see Tables 4.1 and 4.2, respectively. For reasons of simplicity, more encompassing instruments are preferred over regulations per class of technologies, and these latter over the regulation of some specific technologies. The instruments in columns on the left have preference over those in the middle columns, and those in the middle over the instruments in columns on the right. Taxes are more broadly applicable than subsidies, and both are more flexible and incentive-oriented than regulations. As a result, the upper left cell is first choice, with other instruments following only where necessary.

A general fossils input tax on the basis of CO_2 is the simplest instrument for administration purposes in the energy sector. The basic reason *not* to use this instrument is that it does not belong to the Planning & Control political philosophy, even less so than the fossils certificate system.

The second, also flexible, incentivizing encompassing option reverses the tax on emissions and turns it into a subsidy for not emitting. In principle, this is economically equivalent to the emission tax according to the Coase theorem (Coase, 1960), but it violates the "polluter pays principle." One major

Table 4.1 Instruments in the energy sector regulating emissions directly

Direct on emissions	*All emitting technologies*	*Class of emitting technologies*	*Specific emitting technology*
Type 5D taxes, in any form	–General fossil energy input tax on CO_2 basis –Encompassing emission tax energy sector	–Emission tax per class of power stations, per kWh, not heat and co-products –Emission tax on refinery emissions	–Emissions tax on one specific sort of installation
Type 5D subsidies, in any form	–Subsidy on emission reduction relative to firms' performance previous year –Subsidy on any carbon storage	–Subsidy on reducing emissions relative to class average previous year	—
Type 6D prohibitions	–Emission standards, as permits per emitter –Emission permits made tradable in full sector; minus biofuels, with emission rights earned on CCS	–Emission standards in power production –Tradable emission rights in electricity production; refineries separately –Obligatory share of CCS	–Installation-specific emission permits *(non-tradable)*
Type 6D prescriptions	—	—	—

Table 4.2 Instruments in the energy sector regulating emissions by proxy

Proxy on emissions	*All (non-)emitting technologies*	*Class of (non-)emitting technologies*	*Specific (non-)emitting technology*
Type 5P taxes, in any form	–General energy use tax per kWh on all fossil power, not on nuclear and renewables	–Tax on all electricity production	—
Type 5P subsidies, in any form	–Subsidy on all non-fossil energy –Subsidy on any energy efficiency improvement relative to that producers' performance last year	–Subsidy on specific non-fossil energy classes –Subsidy on higher energy efficiency than last year's technology average	–Subsidy on specific non-fossil energy production facilities or energy storage facility
Type 6P prohibitions	–Phased closing down of all fossil power stations and refineries	–Closing down least efficient coal-fired power stations	–Closing down some of most polluting power stations
Type 6P prescriptions	—	–Combined cycle in power production obligatory –Heat to be used functionally –Obligatory share of non-electricity energy co-produced	–Obligatory delivery of power station waste heat to district heating systems

administrative issue is to establish how large an actual emission reduction is. It must be quantified relative to some other emission. That reference may be defined in very different ways, ranging from the sector average from the previous year to the emissions expected if the measures had not been taken. The most generic reference seems to be the emissions of each emitter over the previous year, covering all emissions of the sector. For reasons of symmetry, increases in emissions should be taxed. Otherwise, it would be attractive to acquire emitting activities of other firms, with the increased emissions going for free. Splitting off emitting activities in the following year would lead to entitlement to a subsidy for reduced emissions, creating a continuous source of subsidy without emission reduction. New producers pay the tax on all of their emissions. The instrument subsidizes net emission reduction instead of generating tax revenues, as the mirrored tax option would do. The lower energy prices induced by subsidies would increase total energy demand (Bhandari, Giacomoni, Wollenberg, & Wilson, 2017, p. 21). Combined tax-subsidy instruments are also considered in electricity production. This is basically an emission tax paid back as a subsidy to the non-emitting or low-emitting resource, such as nuclear, hydro, new renewables and natural gas (Galinato & Yoder, 2010). Galinato and Yoder estimate that this combination would have only a third of the welfare effect of a somewhat higher tax alone.

The third encompassing option is that of auctioned, tradable emission permits: cap-and-trade. This option is administratively even more complex than the subsidy-with-tax instrument. Auctioned permits act like a tax, though with a highly fluctuating level, blurring the long-term price signal in comparison with the emission tax. Auctioned permits may be applied upstream, not at the actual emitters.

What all three encompassing pricing instruments have in common is that they lend themselves to expansion to the other sectors. If natural gas emissions are taxed in the energy sector, why not tax the natural gas emissions produced by industry and households, and by transport where natural gas is increasingly used? And why not also expand to coal and oil emissions? For reasons of simplicity, not ideology, instrumentation would shift to the encompassing upstream emission tax system as the simplest and most effective emission pricing scheme. Planning & Control does not support this most simple and efficient pricing instrument for reasons of political philosophy.

With financial instruments all tending toward more Institutionalist-type economy-wide instrumentation, only emission standards remain for generic direct regulation. Generic standards may cover all CO_2 emissions related to electricity generation, expanded to other energy carriers such as heat, hydrogen and ammonia. The increasingly stringent standard applies to fossil-based production only, setting the maximum amount of CO_2 per kWh of delivered energy. The varying quality of heat, with a higher or lower temperature, must be dealt with. Such emission standards are not applicable to refineries, which produce very diverse carbon-containing energy outputs and a substantial amount of chemicals. Setting the emission standard per unit of value added could cover power production and refineries alike. Upstream and downstream vertical integration would, however, reduce the emission score without real reductions. The administrative burden is heavy due to highly instable quantities and prices of inputs and outputs.

Other options for direct instruments focus on specific classes of technologies, reflected in the middle column of Table 4.1. Examples are emission taxes and emission standards for specific types of power stations, for refineries and for all more mixed options producing other energy carriers and several non-energy co-products. The emission reduction may be specified per class with permit prices unequal between classes, which leads to higher administrative complexity and higher costs of emission reduction. Would the planning and control aspect be a good reason for such options? Covering all emissions in the energy sector together and equally has clear advantages. Disregarding sector boundaries so as to cover all sectors has the advantage of simplicity, again, in terms of sparseness, completeness, non-overlap and equality of application, and further reduces the overall costs of emission reduction by creating dynamics for emission reduction. However, all such reasoning leads to financial instruments of the Institutionalist type.

Proxy instruments for emission reduction in the energy sector

Planning & Control then moves to regulating proxy variables for emission reduction; see Table 4.2. That is the main direction in current climate policy instrumentation. The emphasis is on bringing non-fossil energy into the market and taking the highest-emitting fossil production out of the market, possibly combined with emission standards as a direct instrument. Preferred measures are those with the most direct correspondence to the non-fossil reality to be created: subsidizing specific technologies at specific locations, such as an offshore wind park or a particular nuclear installation, and closing down specific fossil fuel producers, such as the highest-emitting coal-fired power station or the worst ultra-heavy oil producer. The most-preferred options are shown in the left-hand column in Table 4.2 and the least-preferred options in the right-hand column.

The rising general fossil energy input tax is the most encompassing instrument – type 5P. It does not differentiate among coal, oil and natural gas, leading to indiscriminate energy reduction, only. For long-term halving of energy use and emissions, all energy prices would roughly have to be doubled (Grubb et al., 2018), still far from the 95% reduction target. The Coase-type mirror instrument is a rising subsidy on all non-fossil energy production, roughly the same as the tax per unit of energy, designed to create the price differential required for the widespread shift to non-fossil fuel. Although starting with modest total expenditures, the rising non-fossil share leads to very substantial subsidy volumes and hence taxes for financing them. The highly subsidized energy system would lead to high total energy use. The third generic instrument covering all fossil emissions is the phased closing down of all fossil power stations and refineries, type 6P, accompanied by the same phased reduction of imports of fossil energy products, such as electricity, kerosene, gasoline, diesel, fuel oil, butane, propane and some precursors. It would work, if implemented fully. With supply artificially reduced, prices would rise, giving windfall profits to some. These profits are more extreme than with cap-and-trade systems because part of that profit is in the auction price paid to governments as a type of tax. Neither of these generic proxy instruments seems capable of reducing emissions deeply, nor are they feasible politically as they create high costs with insufficient emission reduction.

Differentiating the proxy instruments between classes of emitters in the energy sector is the next option, represented by the middle column in Table 4.2. Financial instruments, type 5P, such as taxes on electricity and natural gas, are widely used. They are applied in the use sectors and differentiated between users, covering both the industrial sector and the buildings sector. In the energy sector they would be applied on the inputs to refineries and power stations, with no good reason to differentiate the tax on crude oil emission of refineries or power stations. Similar reasoning applies to all the sectors: why differentiate the tax on natural gas input between use in power stations, in fertilizer production and in heating for industry and buildings? The same reasoning holds true for crude oil and its products. An output tax, for example, on electricity, favors natural gas and hydrogen use unduly and reduces the attractiveness of electric vehicles in transport and heat pumps in buildings. If it covers fossil-based electricity only – a difficult issue in mixed installations – it becomes an input tax. In this scenario, Planning & Control reasoning first considers a generic fossil input tax, and then goes on to specify that tax based on carbon content, not energy content.

Coase-type mirroring again creates other options, such as generic subsidies for specific classes of renewable and nuclear energy. Feed-in tariffs for PVs are a production subsidy funded from a tax on electricity sales. Elegant subsidy systems have developed for larger producers, such as in wind turbine parks. Investment lots are specified with a bidding procedure on the level of a price subsidy; the bidder asking for the lowest subsidy receives the lot. Capital subsidies, often provided as tax deductions, are widely used, and require other taxes for their funding. Further subsidies relate to free provision of grid connections and, more indirectly, access to wholesale electricity markets without payment for backup capacity, with the costs paid by others.

Direct prohibitions and prescriptions, type 6P, first concern the forced closing down of coal-fired power stations, beginning with the worst emitters. Emission standards, type 6D, then would be a preferred instrument, making closing down a private decision. Technical prescriptions relate to energy efficiency, as in prescribing combined cycle heat and power plants and prescribing a minimum share of co-produced non-electricity products, such as heat and hydrogen. This instrument creates advantages for such energy-using systems in the other three sectors, for example, in district heating, hydrogen fuel cell drives and hydrogen aviation. These would compete with independent developments of other technologies, working as subsidies funded from higher electricity prices that others would have to pay. A modest option would be to make heat delivered to district heating systems obligatory if there is a connection to a power station.

The extensive use of subsidies has two disadvantages: they require funding, and they lead to higher energy use and hence lower energy efficiency in society. "Permanent renewable energy subsidies, however, are a very poor and risky substitute for missing carbon prices in the long-run. Mitigation costs increase by a multiple if no carbon price is available, for variations in a wide range of plausible parameters" (Kalkuhl, Edenhofer, & Lessmann, 2013, p. 220). Combining energy subsidies with an energy tax would at least partially resolve these disadvantages. That energy tax must not be symmetrical to the subsidy; otherwise it will cancel out the regulatory effect. Higher electricity prices relative to natural gas are, however, detrimental to heat pump development.

Exemplary instrumentation for nuclear energy

Nuclear energy is disputed mainly for non-climate reasons and costs. Germany is closing down its nuclear power stations while a comparable post-industrialized country, Great Britain, is building and planning new ones. No stance is taken here. A significant number of nuclear power stations, around 225, is under construction or planned, many in China. These include novel technology options such as small-scale uranium-based power stations; thorium-based options and mixed systems are also being considered (Ion, 2017; Zhou, Yang, Park, & Wu, 2018). Direct co-production of hydrogen is an option, with experiments ongoing in the Soviet Union and China (Ping, Laijun, Songzhe, & Jingming, 2017), as is use of heat (Dincer & Acar, 2015). Nuclear energy can be made competitive by subsidies or by raising the price of fossil competitors, such as by a generic emission tax or similar pricing. That option does not, however, belong to the Planning & Control instrumentation repertoire. Therefore, only proxy-variable instruments for inducing nuclear power production can be considered, and then, only the subsidy instrument – type 5P – seems relevant. Subsidies include a price subsidy or guarantee per kWh, several forms of a capital subsidy, some form of subsidy on future waste-processing, and restricted accident liability. Such subsidy schemes induce full-capacity operation of the facility during the very long subsidy period. With a decreasing role for fossil fuels, wholesale prices will be deeply depressed during a substantial part of nuclear operating times, when the wind blows and the sun shines, requiring high subsidies for nuclear energy. An additional subsidy on dispatchability would create an incentive for production variation, such as through storage, and for co-production of non-electricity products. Such a subsidy will likely be difficult to design. Nuclear subsidies are simple in their instrumentation, but costly in terms of the total of amount of subsidy. Nuclear innovation is also substantially based on public funding, as technical, economic and regulatory uncertainty and the lengthy development period make private funding unattractive.

Main instrumentation options in the energy sector

There are two main options for *direct* emission-oriented instruments. The first instrument is the subsidy for emission reduction, type 5D: it is costly, but also capable of creating some incentives for innovation. The second option is setting emission standards per kWh of electricity, type 6D. Increased stringency will lead to closing down the highest emitters and highest-cost installations first. Setting such standards for groups of installations may reduce the costs of early close-down somewhat. Standards may induce CCS if adequate rules are incorporated, and may also be designed to incentivize co-products.

Primary instruments for regulating *proxy* variables are subsidies for non-fossil energy production – type 5P. Production subsides tend to fragmentation of the electricity market, making subsides on capital investment preferable. Substantial dispatchable capacity for primary and secondary energy must be created, including by type 5P subsidies, for solar and wind (which are variable, intermittent renewables and are non-dispatchable), and for nuclear energy (which is difficult to make dispatchable). There is limited real-time price variability in fragmented electricity markets with fixed prices for different producers and users, requiring substantial capacity payment for dispatchable production, per kWh and kW, for grid stabilization. Finally, some type 6P proxy instruments may regulate key technologies in a binding way, inducing short- and medium-term energy efficiency increases, as by prescribing combined cycle electricity with combined heat and power (CHP) production for district heating and co-produced non-electricity energy carriers such as hydrogen. This may create a long-term lock-in to such fossil systems, at variance with the 2-degree/1.5 degree climate goal. Phased close-down of the highest emitting installations is better carried out with increasingly stringent emission standards. See the first survey of main options in Table 4.3.

Table 4.3 Survey of feasible Planning & Control instrumentation for the energy sector

Type 5 direct	Generic subsidy on CO_2 emission reduction
Type 6 direct	CO_2 emission standards on fossil power stations, covering all energy outputs
Type 5 proxy	Subsidies on all non-fossil energy production
Type 6 proxy	Efficiency regulations forcing in non-electricity energy products

Occam's simplicity razor leads first to a single, direct climate policy instrument – CO_2 emission standards – increasingly stringent, and preferred over the much more complex generic subsidy on CO_2 emission reduction per firm or installation. Subsidies for non-fossil energy production and secondary production are added for reasons of security of supply. Prime public task instruments are subsidies for zero-emission technology development, type 3, and spatial planning to accommodate new energy technologies, type 2.

Instruments for a zero-emission transport sector

The transport sector surveyed

Transport is currently fully dominated by fossil fuel drives, which produce around one sixth of global CO_2 emissions and a much higher share in richer countries. Direct transport emissions are determined by fossil fuel use only, CCS not yet being possible. The transport sector is set to grow with rising incomes and with further globalization (Goodwin, 2012; Millard-Ball & Schipper, 2011). Transport modes include person and freight transport on road, rail and water, in air and space, and in pipes and tubes. Nearly complete replacement by non-fossil drives is technically possible, as rockets for satellites have already shown. Additionally, European data show that very rapid shifts in drives are possible; see Figure 4.4. However, non-fossil drives are as yet negligible.

The 40-year target is zero direct emissions. An option which is widely advocated is bioenergy, from wood, oil-containing seeds and many plants, which take their carbon from the air. However, bioenergy must compete with land in terms of food production and nature conservation. For creating negative emission the bioenergy produced must be destined for power stations with CCS, which is not possible for the mobile and small scale emissions from transport. The other current transport instrument is fleet standards, which began with the CAFE standards[3] in the United States (1975). Fleet standards do not relate to the number of kilometers driven, nor to the number of vehicles or driving style. Additionally, they demonstrate an increasing and extreme discrepancy between standard emissions as recorded in a lab and real-life emissions (Tietge, Mock, German, Bandivadekar, & Ligterink, 2017). They are extremely costly and complex (Anderson, Parry, Sallee, & Fischer, 2011; Millard-Ball & Schipper, 2011; Parry, Evans, & Oates, 2014). Premiums given per electric vehicle sold allow for higher fossil emissions by others. Sales outside the fleet standards domain also rise

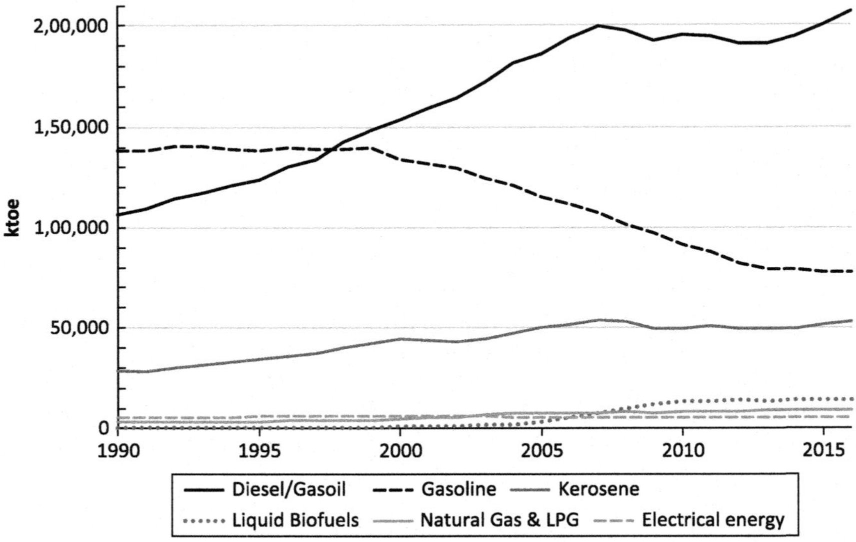

Figure 4.4 Use of fossil fuels in transport in the EU, 1990–2015.

Source: Eurostat[2]

quickly; an example is the small trucks now dominantly being used as personal automobiles in the United States.

Direct instruments for emission reduction in the transport sector

The simplicity criterion leads first to a choice of type 5D, a generic emission tax upstream-administered, as shown in the upper left cell in Table 4.4. A tax of $400 per tonne of CO_2 amounts to $1 per liter of gasoline. For a middle-class passenger car, this is around $1,000 per year and $12,000 over its lifetime.[4] That emission tax may cover fossil fuel use in all other sectors as well, but would then violate the political philosophy and governance ideology of Planning & Control. Another option is a tax levied on each vehicle based on its expected lifetime emissions, shifting the choice of vehicle type to lower emitting and non-fossil. The upfront burdens would be very substantial, certainly for trucks, ships and planes.

The Coase-mirroring of the lifetime taxes, type 5D subsidies, may be given for any combustion fleet lifetime emission reduction of a supplier relative to the previous year's fleet. If the supplier increases expected lifetime emissions of the fleet sold, the subsidy is negative, a tax. Lifetime emissions are specified as the emissions per kilometer multiplied by distance driven per year and the expected lifetime of the vehicle type, averaged for the fleet brought onto the market by a producer. The subsidy level per tonne CO_2 can be similar to an equivalent emission tax. However, unlike the tax, the subsidy would have the effect of lowering user costs and increasing overall transport volumes. Lifetime emission taxes and

Table 4.4 Instruments in the transport sector regulating emissions directly

Direct on emissions	*All emitting technologies*	*Class of emitting technologies*	*Specific emitting technology*
Type 5D taxes, in any form	–General transport fossil fuel tax, CO_2 basis	–Emission fuel tax per transport mode, CO_2 basis –Emission tax on expected lifetime emissions, differentiated per class of vehicles	—
Type 5D subsidies, in any form	–Subsidy on average lifetime emission reduction of any company's combustion vehicle fleet delivered, relative to previous year's fleet	–Subsidy on emission reduction per unit sold of a mode and type relative to a reference type	—
Type 6D prohibitions	–Capped total transport emission rights administered at fossil fuel level, tradable –Vehicle sales permits per amount of lifetime vehicle emissions.	–Emission rights per transport mode and type administered per fossil fuel –Emission standards per transport mode and type, expected yearly/lifetime emissions, per producer fleet	
Type 6D prescriptions	—	—	—

subsidies are substantial in the year of sales. Fuel taxes would be simpler, more effective and more feasible administratively.

Quantity-based instruments, type 6D prohibitions, first may cap the total amount of fossil carbon in transport fuels. The transport fuel certificates can be auctioned. Fuel prices will rise roughly along with the auction price. Administration is at the level of the fuel suppliers, in real time, for example at refinery output for domestic use and imports, similar to current fuel taxes and other duties and excises. This certificate-type instrument can better be expanded to cover all sectors, easing administration, increasing effectiveness, until eventually becoming an economy-wide certificate system. This approach rubs against the control ideology in Planning & Control. The tax variant of pricing is simpler and more effective. The other generic prohibition option is to cap the number of all fossil fuel vehicles, based on measurements of their total lifetime emissions. Flexibility can be created by making the yearly reduced cap tradable, for example, by basing it on emission points. One big truck then may equal 20 small passenger cars. This instrument seems most applicable and effective, even if establishing lifetime emissions is somewhat arbitrary. Producers will check each other. International transport products are more complex, and better handled based on additional international agreements.

Instruments covering emissions per class of technologies are intended to cover all modes, ranging from passenger cars, vans, and small trucks buses to bulk carriers and airplanes – the middle column of Table 4.4. Type 5D as a fuel tax might be differentiated per fuel per transport mode and type. This approach seems to lack any good reasoning, as an equal tax on all carbon in fuels is much simpler. Next is an emission tax on expected lifetime emissions of a vehicle type. It is easy to apply. Quantification is most difficult for vehicles with a long lifespan and possible upgrades, such as trucks, ships and airplanes, which together generate a major and increasing share in transport emissions. The mirrored option, a subsidy on lifetime emission reduction per vehicle type, has the same problems. In highly international aviation, shipping and trucking, the choice of country which is to pay for the subsidy is arbitrary, making this an unlikely option on the basis of national policy alone, while good reasons to choose some country seem difficult to come up with. Application only to nationally registered road vehicles seems possible but then requires other instruments for the other modes of transport.

Prohibitions, type 6D, can be formulated at the fuel level or at the vehicle level. A cap can be set for fuels of different types, such as gasoline, diesel, kerosene, fuel oil and more, linked to modes and types of transport. An overall cap in terms of carbon, as shown in the left-hand column, would be easier and less costly to apply, and create flexibility between modes and types. More aviation transport then implies less use of other fossil fuel–based transport. Differentiated per class, the instrument can function like the multi-sector EU-ETS, which also covers aviation separately. Covering all modes equally would first result in an emission certificate system and, as a better next step, the encompassing emission tax, which is not acceptable in the Planning & Control ideology.

Vehicle emission standards differentiated per transport mode and type – shown as type 6D in the middle column of Table 4.4 – would require an emission specification per vehicle, or at least per vehicle type. A more feasible emission standard specifies the expected lifetime emissions per transport type in each mode. The narrower the boundaries of mode and type, the clearer the control will be. As standards cover all modes and types, it does not so much matter how narrowly the types of fleets are defined as long as all vehicles are covered similarly, from small passenger cars to large trucks, and equivalently for other modes. What is *similar* then has to be defined, related to expected emissions. Although these standards do not regulate the number of combustion vehicles sold, increased stringency will ultimately crowd each type out of the fossil combustion market.

Bringing the number of vehicles sold by a firm under the standard leads to a very different instrument: auctioned sales permits per amount of expected vehicle emissions, made tradable. In this case, a smaller number of higher-emitting vehicles is covered by the same permit amount as a higher number of low-emission vehicles. The volume per year of sales permits then corresponds to the total emission volume that will result until fleet end-of-life. Administration is at the transaction level of sales. Classes of vehicles are used only for specifying expected emissions. This instrument then becomes generic. A fuels-based cap avoids the complexities of establishing lifetime expected emissions per vehicle type. But such a cap can be better transformed into either the generic emission certificate instrument or the generic upstream emission tax, both of which are economy-wide and do not belong to the Planning & Control approach.

Proxy instruments for emission reduction in the transport sector

The upper-left instrument cell in Table 4.5 includes a tax on all fuels or all fossil fuels by mass or energy – type 5P. It is inferior to one on the basis of carbon, a direct instrument, and can best be avoided. Current fuel taxes are also not linked to emissions. Such fuel taxes may easily be transformed to a carbon based tax,

Table 4.5 Instruments in the transport sector regulating emissions by proxy

Proxy for emissions	*All emitting technologies*	*Class of emitting technologies*	*Specific emitting technology*
Type 5P taxes, in any form	–Energy tax on all fuels, all fossil fuels –Tax on all combustion transport (per value, power, mass, person-km, tonne-km, etc.)	–Road pricing per vehicle type –Tax on combustion vehicles (per mode, on value, power, mass, person-km, tonne-km, etc.)	—
Type 5P subsidies, in any form	–Subsidy on noncombustion fuels –Subsidy on noncombustion drive vehicles (on value, power, mass, etc.)	–Subsidy on noncombustion fuels, per fuel type –Subsidy on noncombustion drive vehicles (per mode and class, on value, mass, power, etc.) –Subsidy on demolition of fossil fuel combustion vehicles	—
Type 6P prohibitions	—	–Fossil drives vehicle rationing, per mode and type (number of license plates)	–Extreme emitters taken off road, water and air
Type 6P prescriptions	—	–Prescribe share of non-fossil transport energy for energy use per mode and type of transport vehicle	—

type 5D, possibly applied in all other sectors as well. Subsidies for all non-fossil fuels, such as electricity and hydrogen, are implemented to outcompete fossils and will be based on their energy content.

A generic type 6P tax on all fossil transport vehicles, from passenger cars to ships and planes, seems most feasible as a sales tax, not one that covers actual emissions. The subsidy on all non-fossil combustion drive vehicles mirrors such taxes. Both instruments are widely used and easy to apply. A prescription mandating zero-emission vehicles as an increasing share of combustion drives is not possible generically, as a small electric car does not compensate for the emissions of a large truck.

Instruments per class of technologies, the middle column in Table 4.5, may first involve taxes and subsidies on fuel use or on the vehicles themselves, type 5P. Taxes per vehicle-kilometer (e.g., road pricing), person-kilometer and tonne-kilometer are options, levied only on vehicles run on fossils. Better options exist with emission-oriented instruments. Taxing fossil combustion vehicles themselves requires a basis for setting the tax, such as sales value, power, mass, piston content, transporting capacity in persons, or per tonne, per volume, and so on. Such tax bases are already used for non-climate reasons. However, from a climate policy perspective, not relating the vehicle tax to emissions seems like a waste of regulatory effort.

Subsidies for non-fossil drives can make them attractive. The first option is to subsidize their fuels – electricity, hydrogen, or other source – but then preferably set up generically for road, air and marine transport on the basis of their energy content (the left-hand column in Table 4.5), and then directly on the basis of their carbon content, type 5D. Subsidies for zero-emission vehicles per class may be based on the price difference with their fossil equivalents, but this will differ per mode and class and is tedious to calculate compared with the sales value subsidy. Prohibitions and prescriptions, such as rationing (license plate caps) and prescribed shares of zero-emission vehicles, hardly seem feasible generically and are unduly rigid and expensive.

Exemplary instrumentation for aviation

Aviation emissions are rising rapidly. Aviation is technically and administratively the most demanding mode of transport due to extreme international competition and the current lack of non-fossil technologies. Production is oligopolistic at a global level now, dominated by Boeing and Airbus. Current tax exemptions and reductions under International Civil Aviation Organization (ICAO) agreements[5] must be abolished, a UN action that is long overdue. Policy instruments must create a substantial price advantage for zero-emission flight. Fuel costs are now the completely dominant cost factor. Non-fossil flight is technically close to being implemented, with hydrogen flight as one main option. Small planes running on electricity are already being used commercially for very short flights. Energy density of hydrogen is close to three times that of kerosene,[6] but its use still requires heavy on-board containers, with developments to lighter ones ongoing. (Bicer & Dincer, 2017; Stadler, Caliandro, Leonardi, & Van Herle, 2014). The pricing instruments used first are emission taxes on kerosene. Subsidies for zero-emission fuels are the second choice. Unilateral national instruments will play a role initially, possibly in larger countries only – China, India, the United States and the EU – with taxes on non-climate–policy countries whose planes land there. Due to the long lifespan of airplanes, well beyond 2060 for currently acquired planes, the penetration of zero-emission planes must be forced in sooner, first based on the higher price for kerosene, and subsequently with (international?) subsidies to encourage scrapping of high-emitting planes.

Main instrumentation options in the transport sector

To reduce emissions, the number of all fossil fuel–based vehicles must gradually be reduced to near zero by 2060. This includes shipping and aviation. Inducing efficiency increases in fossil drives exacerbates lock-ins that currently exist. Non-fossil transport must at some point outcompete fossil transports. Implementing generic emission pricing systems would be a prime choice but would violate Planning & Control principles.

Occam's razor then leads to a single regulatory instrument: standards on lifetime-expected emissions per detailed vehicle type that cover all fossil-drive vehicles per firm. These standards, as permits, are made tradable over all transport sectors, including shipping and aviation. Subsidies for zero-emission fuels, as in shifting taxes to fossil fuels, and subsidies for zero-emission vehicles might be added to induce a faster reduction, but only if the standards then are lowered accordingly, which seems impossible from an administrative standpoint. Table 4.6 summarizes these options for the transport sector. In aviation, substantial innovation is first necessary, most likely aimed at perfecting a hydrogen drive. Public task instruments to achieve these goals will involve innovation, infrastructure and spatial planning, as outlined in later sections.

Instruments for a zero-emission buildings sector

The buildings sector surveyed

The share in total direct emissions for heating in buildings is between 10% and 15% in developed countries,[7] and somewhat less in developing countries (Nejat,

Table 4.6 Survey of feasible Planning & Control instrumentation for the transport sector

Type 5 direct	Generic emission pricing on fuels (not acceptable in Planning & Control)
Type 6 direct	Standards on lifetime fleet emissions per detailed type, broadly tradable
Type 5 proxy	Subsidies on zero-emission drive fuels and vehicles (conflict with standards)
Type 6 proxy	Efficiency increases in fossil drives (not desirable, as it creates lock-ins)

Jomehzadeh, Taheri, Gohari, & Abd. Majid, 2015) due to their use of biomass. Energy use in buildings is dominated by space heating and cooling and hot water use, followed by kitchen heating and, mostly electric, lighting and appliances use. Upstream emissions are much higher as virtually all electricity, goods and services, including their transport, and the buildings themselves come from other sectors (Druckman & Jackson, 2016). Space heating is required in moderate and cold regions, and to some degree in subtropical areas, with little space heating in the tropics apart from mountainous areas. Next to natural gas heating, other fossils are used, especially where natural gas distribution is lacking, such as heating oil, liquefied petroleum gas (LPG), butane/propane, and coal.

Emission reduction through energy use reduction can play only a limited role on the road to zero direct emissions in this sector. Insulation and heat exchangers on ventilators can reduce energy use, but not to zero, and will induce higher emissions in the other sectors as for their production (Levy, Woo, Tebbens, & Nishioka, 2017). Solar heating can meet a substantial share of heating requirements in moderate and subtropical areas, but not in winter, not in more northern regions, and not in densely built areas or high-rise buildings. Heat pumps with heat storage may also be attractive, possibly combined with solar heat and zero emission district heating. Solar PV panels may be put on roofs of buildings to supply electricity. They do not eliminate the need for grid connection, however, which means that the location of PV panels is not linked to buildings. They are placed preferably where sunlight is greatest, and are cheapest in dry and sparsely populated areas. Zero-emission energy use in buildings thus must involve zero use of fossils, requiring a shift to zero-emission energy carriers – mainly electricity and hydrogen, and perhaps others – produced through emission-free technology. The technical measures required will increase emissions in other sectors in the short term, such as for the technical installations, for insulation materials and for the additional building space required, but these sectors would then move to near-zero emissions, as well.

Direct instruments for emission reduction in the buildings sector

Instruments for emission reduction can be applied directly at the level of fuels or apparatus to shift demand through pricing or standards. Generic application of instruments, shown in the left-hand column of Table 4.7, is preferred over application to classes of emitting technologies.

Emission taxes covering all emitters, type 5D, are a prime choice, with upstream application to natural gas suppliers, refineries and traders across the entire sector. An even simpler approach is to apply the emission tax to other sectors as well, but this goes beyond the domain of Planning & Control ideology.

Generic subsidies for emission reduction covering all emitters, the second version of type 5D, mirrors taxes, the Coase inverse, which violates the polluter pays principle (Coase, 1960; OECD, 1972). The reduction can be established on a yearly basis, relative to emissions in the previous year, and be administered either per source (households, firms, etc.) or upstream at the suppliers of fossils to

Table 4.7 Instruments in the building sector regulating emissions directly

Direct on emissions	*All emitting technologies*	*Class of emitting technologies*	*Specific emitting technology*
Type 5D taxes, in any form	–Emission tax on natural gas and all other fossils based on carbon content	–Emission tax on natural gas, other fossil heating differently –Tax on lifetime emissions of apparatus	—
Type 5D subsidies, in any form	–Subsidy on reduced fossil use emissions relative to previous year, administered at direct emitters or upstream suppliers	–Subsidy on reduced natural gas use emissions relative to previous year, per dwelling. –Similar but more difficult for other fossil fuels	—
Type 6D prohibitions	–Cap on fossils supply CO_2 basis, administered at provider level	–Emission standards per: household, m^2, m^3, person, etc., including district heating –Standard zero: no fossil supply for new buildings and after deep renovations	–Phase-out of fossil supply for some buildings, dwellings, offices, blocks, owners
Type 6D prescriptions	—	—	—

households. Symmetry requires a negative subsidy on emission increase – a tax on that increase. Funding requires some sort of tax, including a natural gas surcharge. Administration is readily available with respect to grid based natural gas, but not for butane, propane and heating oil. This emission reduction option therefore moves to the class of emitting technologies in the middle column of Table 4.7.

An encompassing cap on fossil fuel supply to buildings, a prohibition of type 6D, can be administered at the level of the provider only, creating a market at that level in auctioned fuel emission permits, which is reduced annually. Such a system at the sector level would lead to unnecessary cost and complexity. Expanding it to the full economy is simpler and would be more effective, but again falls outside the domain of Planning & Control.

Emission taxes on classes of technologies, the middle column of Table 4.7, would be applied to fuels used or to apparatus used. A tax on natural gas would be the first step. But why leave out the other main fossils: heating oil, butane, propane, LPG, and even coal? Including them would lead to a generic sector emission tax, with a strong logic for applying the same tax in the other sectors as well, thereby creating an upstream-administered emission tax – an Institutionalist instrument.

Subsidies for emission reduction can be applied to well-administered fossil fuel users, for example natural gas (see previous paragraph), but not to the other heating fuels. These must then be covered by other instruments, an approach that is also necessary where there is no natural gas grid. This is possible for taxes administered upstream, but not for subsidies.

Emission standards per class of emitting technologies, such as households, nurseries and offices, follow. Setting effective standards requires emission measurement per unit, with a reference for the standard as per square or cubic meter, number of rooms or similar. In addition, collective systems and district heating must then be covered. Measuring natural gas use is quite simple, as it is metered at the level of an individual household, but is not possible for other fossils. A strong cold spell may breach the standard, requiring penalty prices to be instituted. Application seems cumbersome for natural gas and unfeasible for other heating fuels.

The final prohibition regarding emissions is the phase-out of all natural gas supply and forbidding of all fossil fuel–based heaters. The phase-out is easiest for new buildings, by barring connection to the grid and forbidding fossil fuel options in the permit and building regulations. Extensive renovation, requiring building permits, may be tackled in the same way. To cover the full building stock, a phase-out procedure will be required, with the highest-emitting buildings taken off the fossil supply first. This can be extremely costly, however – up to $40,000 for a 100 m^2 dwelling in the Netherlands – and then cost becomes a concern, especially for lower income groups. This instrument seems complementary at best.

Proxy instruments for emission reduction in the buildings sector

A large variety of possible instruments, including for reduced energy use, is available in this category. Instruments include proxy taxes and subsidies, type 5P, and proxy standards, in the form of prohibitions and prescriptions, type 6P, applied either generically or for classes of emitters; see Table 4.8. A general tax on all energy supply to buildings on the basis of energy content would reduce energy use, including renewable energy, increasing energy efficiency in buildings. Applying the general energy tax to fossil energy only would induce a shift to non-fossils such as electricity and hydrogen, regardless of their upstream emissions. The use of heat pumps would not grow with the general tax, but it would with the fossil-only energy tax. Differentiation as to carbon content would be a direct instrument, preferred over proxy versions, and then better applied in all sectors equally.

Mirroring the fossil energy tax, Coase-style, is a generic subsidy on zero-emission energy supply for buildings, issued per kWh for any energy carrier, including heat, hydrogen, electricity, and any other source, including rooftop energy. Such a subsidy would create a sort of indefinite feed-in-tariff or and premium generically that would not be differentiated per installation. Such subsidies must be funded by some tax, preferably not on electricity as in the German feed-in tariff system, but instead on fossils or more neutral taxes, such as the value-added tax (VAT) or income tax. A general tax on all appliances that run on fossils

Table 4.8 Instruments in the building sector regulating emissions by proxy

Proxy on emissions	*All emitting technologies*	*Class of emitting technologies*	*Specific emitting technology*
Type 5P taxes, in any form	–General energy (not CO_2) tax on all energy supply to buildings	–Tax on all fossil heaters being installed (or used) –Tax on higher energy use buildings	—
Type 5P subsidies, in any form	–Subsidy on any zero-emission energy supply (hydrogen and electricity, etc.) –Subsidy on all zero-emission systems installed (per capacity, value, etc.)	–Subsidy on • hydrogen supply • electricity supply • heat pump systems • heat storage system • lower energy use buildings • demolition of fossil heating systems	—
Type 6P prohibitions	–Capping total/fossil energy input in households (per person, m^2, m^3, … etc.) similar for commercial buildings	–Forbidding new gas-, oil-, coal-fired heaters; phasing out old fossil-based ones	—
Type 6P prescriptions	–Minimum insulation values –Maximum energy use per some variable to be chosen	–Obligatory low-temperature heating systems –Minimum energy efficiency (input/useful) for heaters	–Refitting all buildings in urban renewal projects according to highest emission standards

can be levied on their sales value, favoring zero-emission capital investment. The mirrored subsidy on all non-fossil systems can also be based on sales value or expected heating contribution. Experience in Northern Ireland, however, reveals that fraudulent use of such subsidies can easily occur.[8]

Standards, as prohibitions, can also be set up generically, such as capping total energy input to dwellings, offices and similar or capping only the fossil inflow. The basis can be energy (kWh), per season or year, or power (kW) in any period of time. The capped quantity can be related to surface area, building volume or number of persons registered there. Overuse is punished by a fine, stimulating better insulation and higher efficiency apparatus, such as heat pumps.

Proxy instruments per class of technologies can come in great variety. A tax on certain fossil energy sources and not others seems a poor choice since better options are available. Similarly, a subsidy on some zero-emission fuels and not others seems a needless waste of administrative effort. Such technology-specific subsidies may be used for technology development, a type 3 instrument, but then only temporarily.

Taxing high-emitting, existing installations such as low-efficiency heaters and badly insulated buildings are options that do not especially favor zero-emission systems and create lock-ins on improved fossil systems with often a long life expectancy. Taxing new installations based on their expected lifetime emissions is extremely difficult. Also, subsidizing zero-emission technologies is better set up generically. A subsidy for demolition of fossil heating systems in existing buildings will also shift the choice to zero-emission systems in the subsector of existing buildings. Reducing energy use in and by buildings through insulation standards does not favor zero-emission energy use, making even a shift to high-investment systems such as heat pumps with heat storage less attractive. Phasing-out fossil energy systems based on their emissions is a direct instrument and not a proxy instrument.

Exemplary instrumentation for heat pumps with heat and cold storage

Electric heat pumps with heat and cold storage can replace fossil fuel–based heating as one option. Their high energy efficiency reduces variable heating costs but requires higher investments than natural gas–based systems. The difference between electricity and natural gas energy prices (per kWh) determines their economic attractiveness, also dependent to a large extent on local conditions. Systems are highly diverse in temperature range, storage capacity, power, collective supply and combinations with other techniques such as solar heating and district heating. This makes technology-specific instruments very difficult to apply. Reasoning toward some specific option is also soft because some non–heat pump systems may become more attractive in specific situations, for example, where geothermal and solar heat are cheaply available. Also, a given system may be used in very different ways, with some people requiring more high-temperature heating with lower-efficiency installations.

Generic emission taxes constitute the first basic instrument. They increase the costs of natural gas relative to electricity, making electric heat pumps more attractive, especially over the long term when electricity moves toward zero emission and the tax on natural gas goes up. Subsidizing electricity only would work but would create an uneven playing field relative to other systems, such as a hydrogen-based ones. Subsidizing any zero-emission heating technology investment would include heat pumps and lower their capital costs. A buildings sector tax on natural gas can be added, and together they would reduce the overall cost of heat pumps relative to natural gas systems. The phase-out of the natural gas supply grid would force in new systems such as heat pumps.

Additional measures can help overcome the barriers to the broad application of newer technologies and reduce subsidy requirements. Prime regulatory instruments are low-temperature heating requirements in building regulations, which would favor heat pumps. High insulation requirements for buildings reduce the cost advantage of heat pumps, however, due to their high investment costs, especially in existing buildings.

Public research and development may reduce system cost, especially regarding heat exchangers and energy storage, such as for example in molten salts, to reduce

Table 4.9 Survey of feasible Planning & Control instrumentation for the building sector

Type 5 direct	Natural gas and other fossil fuels taxed higher, non-fossils subsidized
Type 6 direct	Standards on emissions: not desirable due to lock-in creation
Type 5 proxy	Generic capital subsidies on non-fossil heating systems
Type 6 proxy	Buildings off the natural gas grid, similar for other fossil fuels

storage costs. A substantial role is available for adapted institutions, initially regarding ownership of underground heat and cold storage. Next, with varying electricity prices in a renewables-dominated market, heat and cold can be stored during low-price periods. Overall, the most relevant instruments can be set up in ways that avoid undue technology focus on heat pumps that utilize heat-and-cold storage alone, however attractive they may seem.

Main regulatory instruments in the buildings sector

Generic emission taxes and cap systems tend toward their Institutionalist version – an upstream emission tax applied equally across all sectors. Use of such a tax would reduce sector-specific instrument requirements substantially, especially if combined with real-time, end-user energy markets. Without them, Occam's razor is less sharp here; see the survey in Table 4.9.

Electric heating systems are made economically attractive by lowering the price difference between electricity and natural gas. This is accomplished through a shift in already existing energy taxes that result in higher taxes on natural gas higher, and similarly for heating oil, butane and propane. New buildings are not connected to the natural gas grid, and existing buildings are taken off the grid with subsidies for required adaptations.

Other heating fuels, such as coal, oil, propane and butane, which are dominant in several countries and regions, may be phased out with additional regulatory instruments. Biofuel subsidies, such as those on wood pellet heaters, are to be abandoned as their role, if relevant, is in large-scale applications with CCS. As technology-specific subsidies are difficult to develop and apply in a dynamic context, a generic capital subsidy then is to be preferred. Energy efficiency measures can actually delay the shift to zero-emission systems by creating lock-ins to fossil systems, which are then cheaper.

Instruments for a zero-emission industrial sector

The industrial sector surveyed

The industrial sector constitutes close to 25% of total global warming emissions, mostly as CO_2 from the incineration of fossil fuels. About half the energy use in the industrial sector is based on fossils, dominated by natural gas. One quarter is heat and electricity from the energy sector, with emissions in that sector, and one quarter has a non-fossil biomass background related to paper production, the manufacturing of wood products such as wood pellets for power stations and heating of buildings, and the processing of numerous agricultural products such as cane sugar. Emissions are dominated by the chemicals industry, including the production of nitrogen-based fertilizer. Iron and steel take second place, with other metals following but in much smaller volumes. Refineries have higher emissions than the chemicals industry, but are treated – increasingly arbitrarily – as part of the energy sector. The industrial processes involved are extremely diverse, both between and within specific industries, with substantially diverging emissions for similar technologies and products. For industries supplying products to more downstream processes, as for example aluminum for airplanes, phasing out is possible only if their role is taken over by other products and industries.

Direct instruments for emission reduction in the industrial sector

The most generic incentivizing instrument is a full-sector fossil energy input tax on the basis of CO_2, the upper-left cell in Table 4.10. This instrument can better be expanded economy-wide for several simplicity reasons: fewer instruments, fewer boundary issues, avoiding real-cost emission taxes in other sectors, and equality. However, the link to specific technologies would then be lacking.

Table 4.10 Instruments in the industrial sector for regulating emissions directly

Direct on emissions	*All emitting technologies*	*Class of emitting technologies*	*Specific production technology*
Type 5D taxes, in any form	–General fossil energy input tax on CO_2 basis –Encompassing emission tax for industrial sector	–Emission tax per industry, based on production volume, class definition –Emission tax on worst producers in a technology class, taxation basis	—
Type 5D subsidies, in any form	–Subsidy on emission reduction relative to firms' performance in previous year	–Subsidy on reducing emissions relative to class average previous year, or to emission standard	–Subsidized close-down of some highest emitting installations
Type 6D prohibitions	–Emission permits per emitter, tradable in the full industrial sector	–Tradable capped emission rights per technology or product domain –Emission standards per technology or product domain, BAT related	—
Type 6D prescriptions	—	–Obligatory share of CCS	—

For this ideological reason it cannot form part of Planning & Control sector instrumentation.

The Coase-type mirror of the tax is a subsidy on reduced emissions by any firm, relative to its previous year's performance. Expansion of emissions leads to a negative subsidy – that is, a tax – making startups difficult. Another version takes sector performance related to the previous year's average per unit of value added as a reference. That is the sector emission intensity, the inverse of eco-efficiency (Huppes & Ishikawa, 2005). Performing better than average yields a subsidy based on the lower intensity of the firm multiplied by its value added. A merger or split-up then is subsidy neutral. Symmetry requires payment of a negative subsidy, as a tax, for a higher-than-average emissions intensity relative to the previous year's intensity. Overall, the result is a net subsidy equal to the full-sector emission reduction per year. Both subsidy schemes require measurement at the level of the firm, a proven obstacle in a cap-and-trade system.

Standards in the form of emission permits for all industrial firms capped and made tradable are the quantity approach. Upstream administration at refineries and other fuel suppliers solves the problems of applying cap-and-trade to smaller firms. Such a system is best applied in the other sectors as well – certainly in the energy sector – going upstream to primary producers and importers. It is easily transformed into a full-covering certificates system and then into the much simpler and more effective emission tax, which is not a Planning & Control instrument.

Direct instruments for classes of technologies are diverse. Taxes per class of emitting technology are ultimately required to cover the full sector, requiring clear class definitions, including covering agriculture as an industrial sector class. There seem to be no reason to differentiate as to the level of emission taxes among classes, however; the generic tax, carbon based, would be preferred.

The subsidy scheme, with negative subsidies applied as taxes for laggards, has advantages if differentiated per class, as there are very substantial differences in emission intensity between classes. A clear definition of classes is required, with subsidies granted for improvement over the previous year's class emission intensity. The subsidy total for the class as a whole would be the reduction of emission intensity of that class multiplied by its value added.

Alternatively, the subsidy scheme could use emissions of the best available technology (BAT) as a reference, thus becoming more like a tax, because the average firm will not reach BAT level. This approach would first require well-established BAT specifications for industry classes and their value added, a tedious task. Emissions and value added would need to be established for each firm in order to specify their emission intensity. All BAT standards together would need to cover the full industrial sector; otherwise, it would be possible to move emissions to non-covered firms, giving only the appearance of reducing emissions. Administration of this option seems nearly impossible. BAT-related emission standards may play a role in providing permits to industries, as in the Seville Process, but seem a far cry from the fast, dynamic emission-reduction instrument that is needed.

Setting a CCS target per firm as a share of its emissions is a costly option for decentralized emitters. Having each firm pay for that amount of CCS – with the carbon for CCS taken from power stations in the energy sector – would function as a partial emission tax and would be better set up as such generically.

Proxy instruments for emission reduction in the industrial sector

Proxy instruments relate, again, to taxes and subsidies, type 5P, and standards as prohibitions and prescriptions, type 6P. Sector-wide generic proxy instruments, the left-hand column in Table 4.11, are initially most incentivizing when taxing fossil fuel inputs on energy content. Implicitly, there are different emission prices. Setting the tax according to carbon content and making it equal for all seems more logical for climate reasons than making it a direct instrument, and one that be better applied economy-wide.

A subsidy on zero-emission fuels by energy content is the Coase-type mirror of the fossils tax on energy content. It does not differentiate between coal, oil and natural gas. The level of the subsidy would have to be very high to force

Table 4.11 Instruments in the industrial sector regulating emissions by proxy

Proxy on emissions	*All emitting technologies*	*Class of emitting technologies*	*Specific production technology*
Type 5P taxes, in any form	–Tax per unit of product by value –Tax per unit of energy input	–Tax per unit of product, differentiated per technology-product combination (e.g., tonne of steel versus tonne of car)	—
Type 5P subsidies, in any form	–Subsidy on all non-emitting energy –Subsidy on reduced energy use for all firms relative to previous year	–Subsidy on low-emission technologies and their products –Subsidy on any energy efficiency improvement relative to previous year, per technology average	–Temporary subsidy on some novel low-emission installations
Type 6P prohibitions	—	–Phasing out all high-emitting installations, highest emitting first as per value, tonne, or other variable, with or without financial compensation	–Closing down some specific high-emission installations
Type 6P prescriptions	—	–Phasing in very low-emitting technologies, as per value, tonne or other variable	—

fossils indiscriminately out of the market. A subsidy on reduced fossil (or general?) energy use per firm, relative to the previous year, would be complex, hardly applicable generically, and would result in limited emission reduction.

Technology standards, type 6P, are impossible to set generically due to the diversity of industries.

Proxy instruments per class of technologies start with emission taxes differentiated per class of technologies. This makes no sense. A tax that is higher on products of the higher-emitting industries would have some effect, taking an industry's emission intensity (emissions per unit of value added) as a basis for the tax. An example would be a high tax on steel and fertilizer, which both have a high emission intensity, and a lower tax on cars and furniture, which both have a lower emission intensity. However, improvements within specific industries would not be incentivized; reductions would result only from substitution. Also, because steel is part of cars, a double taxation would result, leading to undue vertical integration to avoid the tax. Such proxy tax systems may create substantial disturbances with limited emission reduction and are better avoided.

A mirrored subsidy on low-emission technologies and their products, relative to their class, requires precise class definitions and measurement of the emissions of all firms, both hardly feasible. Phasing out the worst performer in a class again requires precise class definitions and measurements per firm. Setting standards in permits based on BAT specifications seems somewhat more viable, but hardly creates strong reduction incentives.

The best short-term option is closure of specific high-emitting installations, such as closing an old steel factory. If there is a clear variation within the class of technologies, this can be done more systematically, first by closing the worst installations, which would somehow need to be defined. In most developed countries, this step would involve compensation to the firms involved. Closure due to market forces is generally accepted; closure by government decree most often is not.

Exemplary instrumentation for iron and steel production

A survey of low- to zero-emission iron and steel technologies can be found in the IEA's *Energy Technology Perspectives* (2017, pp. 186–193), and on the Strategic Energy Technologies Information (SETIS) website of the EU.[9] Near zero-emission steel technologies may use hydrogen for reducing iron ore to iron, and electricity for the heating necessary to produce steel. Lower-emitting technologies also exist. Direct reduced iron (DRI) technology produces significantly lower emissions and is already widely used in some countries, for example, India and the United States.[10] However, it produces a different quality of iron. Shifting from batch processes to continuous flow processes may reduce emissions further, as in the experimental HIsarna-type process, but it does not bring them close to zero. Low-emission technologies take a long time to become dominant due to the long lifespan of installations. Introducing DRI and HIsarna processes now may well delay zero-emission options for decades. Hydrogen and electricity production for zero-emission iron and steel is to be zero emission.

Which instruments would promote a changeover to zero-emission iron and steel, achieved, for example, through use of hydrogen and electricity alone? Generic emission taxes can play a leading role, or industrial sector emission taxes, or emission taxes imposed specifically on iron and steel production. The induced incentive mechanisms are the same, but the administrative applications differ. Subsidized yearly emission reductions, in small steps, will not work to achieve zero-emission technologies as these will require fully new installations. Barring emission taxes, a full iron and steel sector tradable emission cap with auctioned permits could work, similar to the current application in the EU-ETS. From an administrative and political perspective, directly phasing out current technologies seems more of a dead-end than a fork in the road. Public co-funding of technology development such as is now being applied to stimulate the development of hydrogen-based processes, a type 3 instrument, will precede broad introduction of such new technologies.

Main instrumentation in the industrial sector

Emitting technologies are most diverse in the industrial sector. Applying generic emission taxes in only one sector does not make sense; moreover, generic emission taxes are not acceptable in the Planning & Control domain. Emission reduction subsidies are not applicable administratively to small- and medium-sized firms. Subsidies based on exceeding emission reduction relative to dynamic BAT specifications are difficult to design and would have to cover the industrial sector systematically.

Occam's razor is not very sharp here; see Table 4.12. Emission standards per unit of value added for specific high-emitting, larger firms seem the most feasible approach but are only a partial option, applicable where emission measurement per firm is possible, similar to cap-and-trade. The amount per firm is based on its production volume, in terms of value added, multiplied by the standard.

However, for installations with a long lifespan –such as in iron and steel production, where deep innovations and reinvestments are needed – tradable standards do not work. RD&D subsidies are then necessary to stimulate technological development, as long-term risky innovations are unattractive to private investors. Public subsidization of innovation must be followed by forcing or incentivizing instruments for the new zero-emission technologies. Subsidies for zero-emission technology investment can also play a role, overlapping with other instruments.

Table 4.12 Survey of feasible Planning & Control instrumentation for the industrial sector

Type 5 direct	Generic emission taxes (probably not acceptable for ideological reasons)
Type 6 direct	Emission standards for some specific technology groups made tradable
Type 5 proxy	Subsidies on zero-emission technology investment
	Price differential between fossil and non-fossil, same as in buildings heating
Type 6 proxy	Some BAT-based standards for non-covered industries

The price differential between fossil and non-fossil energy carriers, based on shifting diverse taxes, as applied in the buildings sector, can be extended to cover the industrial sector, too.

The overall conclusion is that instruments for fast and deep emission reduction in the industrial sector are much more difficult to design than those in the other three, more homogeneous, sectors. Simplicity reasoning points to a generic upstream emission tax; however, that is opposed to the Planning & Control ideology.

Guide for design of public task instruments

When regulatory instruments of types 4, 5 and 6 do not suffice, higher level instruments are needed: subsidized innovation, type 3; infrastructure and standardization, type 2; and institutions, type 1. Consideration of these additional instruments may start at the sector level, but their impact is, for the most part, more broadly felt. With the growing presence of intermittent wind and solar electricity, flexible energy storage is relevant in all sectors. Hydrogen may become an essential energy carrier not only in mobile applications, but also in households, in industry and in the energy sector itself. An advantage is its ability to accommodate seasonal variation in supply and demand of energy. RD&D for low-emission systems must be considerable given the limited innovation incentives in Planning & Control policy instruments. Research must be diverse, ranging from molecular analysis to development of hydrogen airplanes. New zero-emission technologies also require infrastructure for energy transport and storage, and standardization in order to connect new systems. All of these innovations involve public tasks and also public–private cooperation at least in the buildup stage. Institutions will have to adapt for new markets for electricity, hydrogen and more, and for the growing number of public–private partnerships. Finally, national policies will have to deal with effects of climate regulation on international competitiveness, requiring international cooperation.

Innovation: non-fossil resources, energy storage, long-lasting capital goods

Innovation is necessary where sufficient private funding is lacking owing to absence of incentives and high risks of long–term and uncertain development. Areas of interest in the energy sector include large-scale, zero-emission, low-cost energy storage options and low-priced, zero-emission hydrogen production. In transport, key subjects include high-energy-density storage options for electricity and hydrogen, and zero-emission airplanes. A key issue in buildings is the integration of different systems for heating, cooling, ventilation, and heat and cold storage at different scale levels, up to district heating. Demonstration is the main task, as the complexity of integration is substantial, regarding not just technical details but also management. Novel heating and cold storage options will also first require public funding to demonstrate utility and to create learning curves. In industry, there are few clear focal points due to diversity of technologies. Replacing natural gas–based heating processes with electric heating may be relevant in several industries. Zero-emission iron and steel production requires substantial development and demonstration, with hydrogen-based systems probably being a main line of subsidized development. Other metals production and nitrogen fertilizer also require novel processes to achieve zero emissions. Overall, a substantial amount of innovation will need to be subsidized.

Infrastructure for zero-emission energy

Infrastructure development and standardization, type 2 instruments, concern three main subjects: long-distance electricity transport, hydrogen transport and storage, and modal shifts in transport.

Long-distance electricity transport must link cheap renewables production to cheap storage locations and high end-use regions, and also link supply to different time zones and peak times. Technologies have been well-developed over the past decades, such as in high-voltage direct current (HVDC) systems. Overall, grid design is a public task, with an international component even in large countries such as China and the United States. Implementation can be public or effected in public–private partnerships that create regulated private monopolies. Normal planning procedures are suitable and already generally available.

Hydrogen storage and transport infrastructure seem essential features of zero-emission society, which also cover seasonal variations in supply and demand. Expanding existing infrastructure is an intermediate task for moving toward a full-fledged, all-encompassing system. The public role depends on how markets are developed. Variants of public–private partnerships are possible. Hydrogen supply systems for transport are already being developed with public subsidies, where ultimately private supply may become dominant, as is the case now with gasoline.

Modal shifts toward low-speed transport require a change in infrastructure, including the structure of the built environment (Rauland & Newman, 2015). Effects of measures may be complex, however, as building infrastructure requires materials with high emissions. Introducing high-speed rail and light rail transport, for example, may have deep emission consequences for the infrastructure development required (Chester, Pincetl, Elizabeth, Eisenstein, & Matute, 2013; Eckelman, 2013).

Institutions: energy markets and public–private partnership rules

Some electricity market development is required to accommodate the rising share of non-dispatchable intermittent wind and solar electricity. In electricity markets fragmented by specific taxes, subsidies and standards, some variable pricing needs to be developed. Smaller, more technical items need to be considered, as well. Rules on ownership and on the use of underground heat and cold storage are essential for the broad implementation and use of heat pumps with heat and cold storage. Such rules are largely dependent on national legal systems, and may also depend on the local underground geology, particularly if geothermal heat is an option for use as a heat source.

Variable-price electricity market

Non-dispatchable intermittent and variable renewables, combined with difficult-to-predict demand variations, create disparities in the supply and demand of electricity that need to be equalized by other dispatchable production and storage, and by shifting final demand to better match supply. Electricity markets tend to be fragmented by national subsidies, taxes, price regulations and technical prescriptions. Clear price signals may become opaque or completely absent, providing little incentive for private parties to make relevant and necessary adjustments. Wholesale markets have some variable-priced markets in production now, with renewables not taking part as their marginal costs of production are near zero; they deliver what they can produce as long as the price is positive. Incentives to invest in storage and to use battery systems or hydrogen fuel cells for grid stabilization are lacking for the most part, and must be based on a price differential between storage and secondary delivery. Some positive market development is needed to create relevant incentives.

One option is to allow larger demand organizations, the demand ***integrators***, to become partners on the wholesale market. Only part of the price differential then accrues to the end users and the real-time reaction options for grid stabilization do not materialize. Next, smart meters with some price variation can be prescribed, also based on discrete time periods with fixed prices per period, as in time-of-use pricing. Establishing a real-time variable electricity market with equal prices for all primary and secondary producers and users is hardly possible within the constraints of Planning & Control instrumentation. The substantial advantages of real-time demand adjustment for load management can only be partly realized (Mohsenian-Rad & Leon-Garcia, 2010; Wang et al., 2015; Yi et al., 2013). Time-of-use rates have a fraction of the efficiency advantages obtained with real-time pricing, however (Hogan, 2014).

Reorganizing public–private partnerships

Public–private partnerships are required for dynamics and for sharing risks in domains covered by instruments with the limited incentives of Planning & Control (Osei-Kyei & Chan, 2015), and are also required for bringing in technical knowledge and organizational capacity. Contracts for transparent procurement are clear only if a specific product is delivered, such as a road segment, a bridge, or a cable. But in dynamically evolving energy supply and use systems that are substantially based on dynamically evolving public policies, no single clear product can be defined in advance. Previously in the United States, regulators of the electricity supply determined what was delivered and what the fixed profit share was, effectively reducing incentives for investment, and requiring detailed controls on costs and income. Not fixing the profit rate would similarly have the effect of creating contractually monopolistic firms, where the private partner receives a significant public budget for producing little output. In privately owned, monopolistic transmission grids such as the UK National Grid, underinvestment can still occur, with new investment having to be funded publicly to compensate for a lack of investment in transmission infrastructure. Investment in smart meters to help develop demand response will have to be based on public rules and subsidies regarding partners in the production and retail markets for energy carriers.

After privatization and new public rules for the unbundling of production, transmission and distribution, such as in the EU (European Comission, 2009, 2013), new versions of public–private partnerships must be developed to compensate for market deficiencies created by Planning & Control climate and energy policy instrumentation, including income-compensating measures through differentiated user prices. There are no clear rules on what constitutes good options, with empirical results diverging substantially regarding effectiveness and efficiency (Hodge & Greve, 2007; Hodge & Greve, 2009).[11] Price guarantees and price subsidies for wind, solar and nuclear energy shift market risks to governments, leaving only technical risks for the private firms involved. Explicitly sharing market risks and reducing duration of commitments are two directions to be developed, with transparent rules on ownership, pricing and bankruptcy procedures. Such conditions and procedures link deeply to both public and private law, in complex ways.

International instruments on competitiveness, CCS and energy markets

Competitiveness and carbon leakage

All prohibitions, prescriptions, subsidies and taxes can significantly influence prices. They will change relative prices against those abroad, shifting comparative advantages and hence trade patterns. In a country with a comprehensive climate policy, products with induced higher prices will receive more competition from abroad and may lose out on the international market. However, the country's other products will improve their competitiveness relative to products abroad. This mechanism occurs more diffusely, through increased exports to countries that are less active in climate policy. This has been a basic mechanism behind international trade development, as formulated by Ricardo in the theory of comparative advantage in 1817 already (Ricardo, 2017). Flexible exchange rates accommodate changes in absolute costs, which themselves are not inherent. In a single country with an effective climate policy, some firms may have to reduce production and possibly close down while others flourish. Climate policy will play a role, but mostly not a dominant one in comparison to other factors.

What should be done in terms of international instrumentation? A first step would be to apply internationally the same instruments that are applied nationally. Such international agreements are hardly possible, however, if each country acts according to its own industrial capacities, resources and other endowments to move toward its reduction targets. Some countries will act to limit specific emission sources and not others. The non-acting countries should ideally pay a tax on imports to more active climate policy countries amounting to the additional costs that would be induced if they implemented similar climate policies. Such a border tax adjustment would also avoid the export of emissions abroad in the form of carbon leakage (Antimiani, Costantini, Kuik, & Paglialunga, 2016; Böhringer, Carbone, & Rutherford, 2012; Fischer & Salant, 2014). This leakage constitutes a second argument for border tax adjustment next to the economic one of undue competition: the global climate effect. Border tax adjustments (BTAs) are allowable in principle under WTO rules (Tamiotti, 2011). Approximately a quarter of global emissions is embodied in imported products (Davis, Peters, & Caldeira, 2011), making such adjustments a very substantial affair. Serious quantifications for a broader range of products seem far beyond what economic science can supply. Slightly simpler is an "embodied emissions border tax" (Mehling, Das, Droege, Asselt, & Verkuijl, 2017). Product-specific levels are extremely difficult to establish, however, and are not constant over time due to price and technology developments. A less refined approach is to impose a heavy tax on imports from non-policy countries, to punish them for not joining the climate policy club (Nordhaus, 2015). This would solve measurement problems but seems an almost certain way to provoke a global trade war.

Ricardian reasoning and limited empirical research (Böhringer, Balistreri, & Rutherford, 2012) both indicate that the long-term effects of leakage on national income in the policy country are limited. Easing the adaptation process should then become the main policy goal, protecting only the energy and emission-intensive industry that is substantially exposed to foreign competition, and then only for a limited period of time, as for iron and steel, nitrogen fertilizer, and some base chemicals. In an extreme Planning & Control vein, the EU provides extensive details about which industries are endangered by – as yet limited – EU climate policy (European Commission, 2014).[12] Effective implementation has so far proven difficult.

CCS *agreements*

Contrary to other economic activities newly induced by climate policy, capture and storage of carbon creates costs only, without creating any market value. CO_2 capture, from air as DACS or from any process as CCS, entails cost, as do transport and pressurizing CO_2 deep underground. Countries may differ substantially in terms of storage options and their costs. So why would a country choose to incur the costs of capture? The cost of CCS may be reduced over time, but they will remain substantial nonetheless. Storing 10% of current CO_2 emissions of 40 billion tonnes per year at $125 per tonne CO_2 would cost half a trillion dollars annually. Such costs are not as extreme as they sound, constituting approximately one half of 1% of current gross world product, on the order of 2% of global public expenditure. But agreement on the share of costs per country, and on national funding, will be difficult to reach. Only on the basis of binding international agreements would national CCS be justified for an individual country.

International energy markets

International energy markets have been effectively established for all fossil energy products: coal, oil and refined oil products, and natural gas. Hydrogen can be treated similarly to natural gas, transported by pipeline or as a liquid by ship, and stored underground. No policy is needed. Electricity is different as it needs to be used or stored within microseconds of its production. New technologies allow for the rapid reactions required and for low-cost, long-distance transport. HVDC supergrids already connect some regional transmission grids. Expansion of electricity markets internationally has several advantages. Peaks in production and in use can be evened out. Low-cost production areas can be connected to high-use areas. Lowest-cost storage options can be connected to production and use. These technical options have clear economic advantages once design of a supergrid is optimized and connected transmission and distribution grids become the norm. The instruments necessary to help create the potential advantages are standardization of exchange and transformation protocols, including grid codes, pricing and tax rules, and market design, including failure treatment and liabilities. The supergrid is inherently monopolistic and may be used to manipulate prices, which will require detailed safeguards and controls. International instrument development is ongoing globally in the International Organization for Standardization (ISO) and the Institute of Electrical and Electronics Engineers (IEEE). Additional regional and national organizations exist in China, North America and Europe.[13]

Adjoining measures for non-climate reasons

Energy security measures

Contrary to the reasoning underlying IEA climate and energy scenarios, due to effective climate policy, energy security does not seem to be an emerging issue. Rapid transition toward zero-emission technologies reduces the fossil share and will also reduce the options for monopolistic action such as occurs in OPEC. Energy security will rise autonomously. Security will also improve with increased decentralized storage of electricity, hydrogen and heat. Energy security policy should be independent and developed for extreme events only, and not as a main aspect of combined climate and energy policy. A pessimistic view of the potential effectiveness of climate policy would make energy efficiency policy more relevant. But effective policy aimed toward achieving zero emissions is assumed here.

Income distribution measures

Income distribution issues relating to climate change policy focus on energy poverty (Day, Walker, & Simcock, 2016). Costs of climate policy instruments involve both transfer payments, in terms of a variety of subsidies and taxes, and real increases in the cost of energy. For lower-income groups in poor housing situations, life could become more difficult. This is even more so if incomes stagnate or decline, as is now the case, particularly in the developed Anglo-Saxon countries (Alvaredo, Chancel, Piketty, Saez, & Zucman, 2017). Income distribution after taxes and redistribution differs even more per country, indicating the overarching role of national policies. National climate policies may be a very small part, only, of overall structural development in income distribution. It seems that climate policy is not the best-suited realm for developing and implementing income distribution policy, not even in the costly Planning & Control version. Transfer payments in Planning & Control climate policy may be compensated for in tax policy and public expenditure. Real costs are unavoidable. With better incentives for emission reduction they will be lower.

Key points for Planning & Control instrumentation

Supply-driven change can force downstream emission reductions, and demand-driven change can force upstream reductions. Zero-emission hydrogen supply can reduce downstream natural gas emissions, and zero-emission drives in transport will reduce demand from refineries to near zero, resulting in near-zero emissions. Focusing instruments on energy efficiency improvements of fossil fuel systems increases lock-ins to fossils, and creates barriers to future entrance of zero-emission systems. Energy security improves by a shift to renewables and nuclear and should not be viewed as a climate policy issue. Income inequality and poverty is mostly determined by non-climate policy factors related to education, innovation, IT and globalization, with these causes and redistribution remediation being fully dominant over possible climate policy measures.

In all four sectors, administrative-technical simplicity reasoning regarding more technology-specific instruments tends toward ever-more-encompassing systems evolving to emission taxes, ultimately set up most simply and effectively as an economy-wide, upstream-administered emission tax, with few regulatees and covering all emitters. Although superior from an administrative-technical and economic perspective, such a broadly encompassing tax is at odds with the political philosophy and governance ideology of Planning & Control, which favors more volume oriented technology-specific regulations. Closer to Planning & Control is a full systems approach including fossil energy certificates, similar to cap-and-trade but easier to implement and covering all emitters in all sectors. More specific renewables policies would not make sense, however, under such a system at variance with planning and control.

More complex instrumentation therefore results. It uses technology-specific sticks and carrots: standards and subsidies funded with taxes. The energy sector is covered by CO_2 emission standards on fossil power stations covering all energy outputs and by subsidies for all non-fossil energy production. Transport is covered by tradable type-specific lifetime emission standards, combined with raised fuel taxes and subsidies for electricity and zero-emission drives, replacing failing fleet standards and making refineries mostly redundant. The buildings sector is covered by raising taxes on natural gas, subsidies for non-fossil energy, generic capital subsidies for non-fossil heating systems, and stepwise closing of the natural gas grid. The industrial sector has the most diverse and least systematic instrumentation, using emission standards made tradable for some specific technology groups, subsidies for zero-emission technology development and introduction, the fossil–non-fossil price differential, as applies in heating buildings as well, and some BAT-based standards for non–cap-and-trade industries. Electricity markets remain fragmented, with some time-of-use rate variation, only. As incentives for emission reduction are limited, substantial public funding of innovations is required until their effective market introduction, aided by extensive use of public–private partnerships to accelerate their development.

Shifts in competiveness result from non-global climate policy instrumentation, which also lead to carbon leakage. Border tax adjustments are an option to address these issues but not an easy one to implement, because a clear basis for quantification is lacking. This leaves broader international agreements on climate policy instrumentation as a more basic option, along with agreements covering CCS.

Notes

1. Downloadable from https://www.ipcc.ch/report/ar5/wg3/.
2. Available at http://ec.europa.eu/eurostat/statistics-explained/index.php/Oil_and_petroleum_products_-_a_statistical_overview.
3. See https://en.wikipedia.org/wiki/Corporate_average_fuel_economy, with further references there
4. At 15,000 km per car per year and 15 liters per km, this amounts to 1,000 liters per year. At 2.5 tonnes CO_2 per liter multiplied by $400 per tonne CO_2, the tax is $1,000 per driving year.
5. ICAO is a UN specialized agency founded in 1947. See the third edition of Document 8632, ICAO's *Policies on Taxation in the Field of International Air Transport*, issued in 2000. Available at https://www.icao.int/publications/Documents/8632_cons_en.pdf.
6. See https://energy.gov/eere/fuelcells/hydrogen-storage.
7. See https://www.epa.gov/ghgemissions/sources-greenhouse-gas-emissions.
8. See https://en.wikipedia.org/wiki/Renewable_Heat_Incentive_scandal. The subsidy amounted to 500 million pounds on a population of 2 million.
9. Seehttps://setis.ec.europa.eu/setis-reports/setis-magazine/carbon-capture-utilisation-and-storage/zero-emission-steel-plant-of.
10. See https://www.eia.gov/todayinenergy/detail.php?id=34052.
11. One advantage, on paper only, is that borrowing for public–private provision does not add to public debt but to private debt, albeit being more expensive for the taxpayer.
12. Some examples from a very long list: "Dried potatoes in the form of flour, meal, flakes, granules and pellets," "Frozen potatoes, prepared or preserved (including potatoes cooked or partly cooked in oil and then frozen; excluding by vinegar or acetic acid)," as well as "Whey and modified whey in powder, granules or other solid forms, whether or not concentrated or containing added sweetening matter."
13. In China: SAC, the Standardization Administration of the People's Republic of China. In the United States: FERC, the Federal Energy Regulatory Commission, and NERC, North American Electric Reliability Corporation. In Europe: ENTSOE, the European Network of Transmission System Operators for Electricity.

References

Alvaredo, F., Chancel, L., Piketty, T., Saez, E., & Zucman, G. (2017). Global inequality dynamics: new findings from WID. world. *American Economic Review, 107*(5), 404–409.

Anderson, S. T., Parry, I. W., Sallee, J. M., & Fischer, C. (2011). Automobile fuel economy standards: impacts, efficiency, and alternatives. *Review of Environmental Economics and Policy*, 5(1), 89–108. doi: 10.1093/reep/req021

Antimiani, A., Costantini, V., Kuik, O., & Paglialunga, E. (2016). Mitigation of adverse effects on competitiveness and leakage of unilateral EU climate policy: an assessment of policy instruments. *Ecological Economics, 128*, 246–259. doi: 10.1016/j.ecolecon.2016.05.003

Bhandari, V., Giacomoni, A. M., Wollenberg, B. F., & Wilson, E. J. (2017). Interacting policies in power systems: renewable subsidies and a carbon tax. *Electricity Journal, 30*(6), 80–84. doi: 10.1016/j.tej.2017.06.004

Bicer, Y., & Dincer, I. (2017). Life cycle evaluation of hydrogen and other potential fuels for aircrafts. *International Journal of Hydrogen Energy, 42*(16), 10722–10738. doi: 10.1016/j.ijhydene.2016.12.119

Böhringer, C., Balistreri, E. J., & Rutherford, T. F. (2012). The role of border carbon adjustment in unilateral climate policy: overview of an Energy Modeling Forum study (EMF 29). *Energy Economics, 34*, S97–S110. doi: 10.1016/j.eneco.2012.10.003

Böhringer, C., Carbone, J. C., & Rutherford, T. F. (2012). Unilateral climate policy design: efficiency and equity implications of alternative instruments to reduce carbon leakage. *Energy Economics, 34*, S208–S217. doi: 10.1016/j.eneco.2012.09.011

Chester, M., Pincetl, S., Elizabeth, Z., Eisenstein, W., & Matute, J. (2013). Infrastructure and automobile shifts: positioning transit to reduce life-cycle environmental impacts for urban sustainability goals. *Environmental Research Letters*, 8(1), 015041.

Coase, R. H. (1960). The problem of social cost. *Journal of Law & Economics*, 3, 1–44.

Davis, S. J., Peters, G. P., & Caldeira, K. (2011). The supply chain of CO_2 emissions. *Proceedings of the National Academy of Sciences, 108*(45), 18554–18559. doi: 10.1073/pnas.1107409108

Day, R., Walker, G., & Simcock, N. (2016). Conceptualising energy use and energy poverty using a capabilities framework. *Energy Policy*, 93, 255–264.

Dincer, I., & Acar, C. (2015). Review and evaluation of hydrogen production methods for better sustainability. *International Journal of Hydrogen Energy*, 40(34), 11094–11111. doi: 10.1016/j.ijhydene.2014.12.035

Druckman, A., & Jackson, T. (2016). Understanding households as drivers of carbon emissions. In R. Clift & A. Druckman (Eds.), *Taking stock of industrial ecology* (pp. 181–203). Cham, Switzerland: Springer International.

European Commission. (2009). *Directive 2009/72/EC of the European Parliament and of the Council of 13 July 2009 concerning common rules for the internal market in electricity and repealing Directive 2003/54/EC (Text with EEA relevance)*. Retrieved from http://data.europa.eu/eli/dir/2009/72/oj.

European Commission. (2013). *Ownership unbundling. The Commission's practice in assessing the presence of a conflict of interest including in case of financial investors*. Commission Staff Working Document *(2013) 177 final.* Brussels: European Commission.

European Commission. (2014). *Commission Decision of 18 December 2013 amending Decisions 2010/2/EU and 2011/278/EU as regards the sectors and subsectors which are deemed to be exposed to a significant risk of carbon leakage (2014/9/EU) Available at: http://eur-lex.europa.eu/LexUriServ/LexUriServ.do?uri=OJ:L:2014:009:0009:0012:EN:PDF.* Brussels: European Commission.

Eckelman, M. J. (2013). Life cycle assessment in support of sustainable transportation. *Environmental Research Letters*, 8(2), 021004.

Fischer, C., & Salant, S. W. (2014). Limits to limiting greenhouse gases: intertemporal leakage, spatial leakage, and negative leakage. In K. Pittel, R. van der Ploeg, & C. Withagen (Eds.), *Climate Policy and Nonrenewable Resources: The Green Paradox and Beyond*. Cambridge, MA: MIT Press.

Galinato, G. I., & Yoder, J. K. (2010). An integrated tax-subsidy policy for carbon emission reduction. *Resource and Energy Economics*, *32*(3), 310–326. doi: 10.1016/j.reseneeco.2009.10.001

Goodwin, P. (2012). *Peak travel, peak car and the future of mobility. Evidence, unresolved issues, policy implications, and a research agenda.* Paris: OECD/ITF Discussion Paper 2012-13.

Grubb, M., Bashmakov, I., Drummond, P., Myshak, A., Hughes, N., Biancardi, A., ... Lowe, R. (2018). *An exploration of energy cost, ranges, limits and adjustment process. Final report to the Institute for New Economic Thinking*. London. UCL downloadable. https://www.ucl.ac.uk/bartlett/sustainable/sites/bartlett/files/an_exploration_of_energy_cost_ranges_limits_and_adjustment_process_2_0.pdf.

Hodge, G. A., & Greve, C. (2007). Public–private partnerships: an international performance review. *Public Administration Review*, *67*(3), 545–558.

Hodge, G. A., & Greve, C. (2009). PPPs: the passage of time permits a sober reflection. *Economic Affairs*, *29*(1), 33–39. doi:10.1111/j.1468-0270.2009.01864.x

Hogan, W. W. (2014). *Time-of-use rates and real-time prices*. Retrieved from https://sites.hks.harvard.edu/fs/whogan/Hogan_TOU_RTP_Newark_082314.pdf

Höhne, N., Kuramochi, T., Warnecke, C., Röser, F., Fekete, H., Hagemann, M., ... Gonzales, S. (2017). The Paris Agreement: resolving the inconsistency between global goals and national contributions. *Climate Policy*, *17*(1), 16–32. doi: 10.1080/14693062.2016.1218320

Hood, C. (2011). *Summing up the parts. Combining policy instruments for least-cost climate mitigation strategies*. Paris: OECD/IEA.

Huppes, G., & Ishikawa, M. (2005). A framework for quantified eco-efficiency analysis. *Journal of Industrial Ecology*, 9(4), 25–41. doi: 10.1162/108819805775247882

IEA. (2017). *Energy technology perspectives 2017, catalysing energy technology transformations*. Paris: International Energy Agency. doi: 10.1787/energy_tech-2017-en

Ion, S. (2017). Challenges to deployment of twenty-first century nuclear reactor systems. *Proceedings of the Royal Society A: Mathematical, Physical and Engineering Sciences*, *473*(2198), 20160815. doi: 10.1098/rspa.2016.0815

IPCC-AR5-WGIII. (2014). *Mitigation of climate change. Contribution of Working Group III to the Fifth Assessment Report of the Intergovernmental Panel on Climate Change*. Cambridge, England: Cambridge University Press. Retrieved from https://www.ipcc.ch/pdf/assessment-report/ar5/wg3/ipcc_wg3_ar5_frontmatter.pdf

Kalkuhl, M., Edenhofer, O., & Lessmann, K. (2013). Renewable energy subsidies: second-best policy or fatal aberration for mitigation? *Resource and Energy Economics*, *35*(3), 217–234. doi: 10.1016/j.reseneeco.2013.01.002

Levy, J. I., Woo, M. K., Tebbens, R. D., & Nishioka, Y. (2017). Emission payback periods for increased residential insulation using marginal electricity modeling: a life cycle approach. *International Journal of Life Cycle Assessment*, 1–12.

Mehling, M., Das, K., Droege, S., Asselt, H. v., & Verkuijl, C. (2017). *How to design border carbon adjustments that work for the climate: policy brief*. London: Climate Strategies.

Millar, R. J., Fuglestvedt, J. S., Friedlingstein, P., Rogelj, J., Grubb, M. J., Matthews, H. D., ... Allen, M. R. (2017). Emission budgets and pathways consistent with limiting warming to 1.5 °C. *Nature Geoscience, 10*, 741. doi: 10.1038/ngeo3031; https://www.nature.com/articles/ngeo3031#supplementary-information.

Millard-Ball, A., & Schipper, L. (2011). Are we reaching peak travel? Trends in passenger transport in eight industrialized countries. *Transport Reviews, 31*(3), 357–378. doi: 10.1080/01441647.2010.518291

Mohsenian-Rad, A.-H., & Leon-Garcia, A. (2010). Optimal residential load control with price prediction in real-time electricity pricing environments. *IEEE Transactions on Smart Grid, 1*(2), 120–133.

Nejat, P., Jomehzadeh, F., Taheri, M. M., Gohari, M., & Abd. Majid, M. Z. (2015). A global review of energy consumption, CO_2 emissions and policy in the residential sector (with an overview of the top ten CO_2 emitting countries). *Renewable and Sustainable Energy Reviews, 43*, 843–862. doi: 10.1016/j.rser.2014.11.066

Nordhaus, W. (2015). Climate clubs: overcoming free-riding in international climate policy. *American Economic Review, 105*(4), 1339–1370. doi: 10.1257/aer.15000001

OECD. (1972). *Recommendation to the Council on Guiding Principles concerning international economic aspects of environmental policies*. Paris: OECD.

Osei-Kyei, R., & Chan, A. P. (2015). Review of studies on the Critical Success Factors for Public–Private Partnership (PPP) projects from 1990 to 2013. *International Journal of Project Management, 33*(6), 1335–1346.

Parry, I. W. H., Evans, D., & Oates, W. E. (2014). Are energy efficiency standards justified? *Journal of Environmental Economics and Management, 67*(2), 104–125. doi: 10.1016/j.jeem.2013.11.003

Patterson, M. G. (1996). What is energy efficiency?: Concepts, indicators and methodological issues. *Energy Policy, 24*(5), 377–390. doi: 10.1016/0301-4215(96)00017-1

Ping, Z., Laijun, W., Songzhe, C., & Jingming, X. (2017). Progress of nuclear hydrogen production through the iodine–sulfur process in China. *Renewable and Sustainable Energy Reviews, 81*, 1802–1812.

Rauland, V., & Newman, P. (2015). *Decarbonising cities: mainstreaming low carbon urban development*. Cham, Switzerland: Springer International.

Ricardo, D. (2017). On the principles of political economy and taxation: on foreign trade. In J. R. & W. R. (Eds.), *200 Years of Ricardian Trade Theory* (chap. 7, pp. 233–268). Cham, Switzerland: Springer International.

Rockström, J., Gaffney, O., Rogelj, J., Meinshausen, M., Nakicenovic, N., & Schellnhuber, H. J. (2017). A roadmap for rapid decarbonization. Emissions inevitably approach zero with a "carbon law". *Science, 355*(6331), 1269–1271.

Rogelj, J., Schaeffer, M., Friedlingstein, P., Gillett, N. P., Van Vuuren, D. P., Riahi, K., … Knutti, R. (2016). Differences between carbon budget estimates unravelled. *Nature Climate Change*, 6(3), 245–252.

Stadler, P., Caliandro, P., Leonardi, C., & Van Herle, J. (2014). *Cost evaluation of large scale hydrogen production for the aviation industry*. (Master's thesis, EPFL, Lausanne).

Stephenson, A., & MacKay, D. J. C. (2014). *Life cycle impacts of biomass electricity in 2020. Scenarios for assessing the greenhouse gas impacts and energy input requirements of using North American woody biomass for electricity generation in the UK*. (URN 14D/243). London: Department of Energy and Climate. Retrieved from https://www.gov.uk/government/publications/life-cycle-impacts-of-biomass-electricity-in-2020.

Stiglitz, J. E., Sen, A. K., & Fitoussi, J.-P. (2009). *The measurement of economic performance and social progress revisited: reflections and overview*. Paris: OFCE. HAL Id: hal-01069384.

Sunstein, C. R. (2014). *Simpler: the future of government*. New York: Simon and Schuster.

Tamiotti, L. (2011). The legal interface between carbon border measures and trade rules. *Climate Policy*, *11*(5), 1202–1211. doi: 10.1080/14693062.2011.592672

Tietge, U., Mock, P., German, J., Bandivadekar, A., & Ligterink, N. (2017). *From laboratory to road. A 2017 update of official and "real-world" fuel consumption and CO_2 values for passenger cars in Europe*. Berlin: ICCT. Retrieved from https://www.theicct.org/publications/laboratory-road-2017-update

Wang, Q., Zhang, C., Ding, Y., Xydis, G., Wang, J., & Østergaard, J. (2015). Review of real-time electricity markets for integrating distributed energy resources and demand response. *Applied Energy*, *138*, 695–706. doi: 10.1016/j.apenergy.2014.10.048

Yi, D., Pineda, S., Nyeng, P., Ostergaard, J., Larsen, E. M., & Qiuwei, W. (2013). Real-time market concept architecture for EcoGrid EU-2014; a prototype for European smart grids. *IEEE Transactions on Smart Grid*, *4*(4), 2006–2016. doi: 10.1109/TSG.2013.2258048

Zhou, S., Yang, W. S., Park, T., & Wu, H. (2018). Fuel cycle analysis of molten salt reactors based on coupled neutronics and thermal-hydraulics calculations. *Annals of Nuclear Energy*, *114*, 369–383. doi:https://doi.org/10.1016/j.anucene.2017.10.040

5 Designing Institutionalist climate policy instrumentation

Institutionalist design philosophy: incentivizing and enabling

Redesigning the fabric of society in light of climate change

The organizing structure of society, its institutions, defines how societies are functioning. It involves its constitution, setting decision making rules for changing other institutions such as ownership rules, patent law, trade law, family law, research organization and funding, education systems and tax structure. Ownership must be established in every society, in one way or another. Defining ownership also defines what constitutes theft. Liability rules set the boundaries of what one can do with what one owns as well as with what one does not own. The boundaries of what can and cannot be owned privately are generally quite clear. Most material things including land can be owned privately, but not human beings or the seas. Even states are not permitted to own the seas, a rule established in 17th-century Europe with the rise of multilateral international trade. "The Seas Are Free" (*Mare Liberum)* is the title of a book written by Dutch legal scholar Grotius in 1609 (Feenstra, 2009). The open seas constitute a global public good where infringements are defined and regulated even in the absence of a global authority to correct infringements.

A stable climate is a comparable global public good, but it is as yet ill-defined with respect to allowable infringements of the air around us. In contrast to open seas, which could be delineated as seas existing beyond a given distance from land, the air circulates freely around us and around the globe and is in that sense a most pure public good. The skies and space are free for use by airplanes and rockets a certain distance above the ground and fully above the oceans, with only very limited rules having to do with avoiding collisions. Air traffic congestion may lead to changing these rules. Similarly, the air can be used for emissions, with limits based on property and liability rules. The essence of private liability is that one is entitled to be free of infringement by others, even if the other party has clear economic advantages for infringing on what one owns, including one's body and health. Toxic fumes generated by a local neighbor can mostly be dealt with through liability procedures. As the numbers of emitters and injured parties have grown, liability rules have expanded to provide strict and several liability, chain liability and the possibility of prosecuting class actions. That is a first line of institutionalist reasoning.

As the types and sincerity of environmental infringements have increased, theoretical reasoning based primarily on economics and public welfare has supplanted liability reasoning. From this perspective, infringements are tolerated if they create an overall societal advantage. If the injured are fully compensated and a positive value remains for all to enjoy, one can speak of an increase in welfare: the Pareto criterion. The Pareto criterion has been weakened in complex societies to the "potential Pareto improvement," meaning that the injured could be compensated fully, but need not be. If advantages and disadvantages are distributed somewhat randomly throughout the population, on average everyone will have an advantage from larger numbers of potential Pareto improvements. This reasoning forms the basis of applied welfare theory, from cost-benefit analysis to collective welfare functions. It also forms the basis of much reasoning in the climate policy domain, where there is a strong focus on establishing the value of damages caused by emissions. Economists disagree considerably on methods and outcomes (Portney & Weyant, 1999; Tol, 2008) but most agree that these numbers should form the basis for policy and its instrumentation. Broad optimality is to be approached, with active compensation only for those most affected. This economic welfare reasoning forms the basis for Planning & Control instrumentation in climate policy – specifying how society can improve as part of broader governance. This is *not* the reasoning upon which Institutionalist climate policy instrumentation is based.

Not all political economists subscribe to this optimality-based reasoning in climate policy. A few argue that it is not possible to establish damage numbers in a meaningful way for the most serious climate problems that may develop (Weitzman, 2011). Climate change could involve at least regional apocalyptic breakdown of societies, accompanied by hunger, illnesses and war, all linked to ecological disaster. There is a chance greater than zero that runaway mechanisms will jeopardize stable society on a global scale – a low probability but with a very high impact. This is a first line of reasoning away from integrated and mainstreamed climate policy as part of optimal public policy. Society aims to prevent climate disaster, which has potential effects incomparably more destructive than nuclear power disasters.

The second line of reasoning away from optimality has a different starting point altogether: some ownership rights simply cannot be traded against the wishes of those involved. The future is not tradable. Faust should not be our friend. Climate change should be curbed for reasons other than collective welfare, however broadly that may be defined beyond economics (Nussbaum & Sen, 1993). Envisioning a long future in a stable society with happiness for your own kind is an honorable goal, as are other normative ethical views based on moral and religious theories. With applied welfare theory becoming less relevant and the usual liability reasoning becoming inapplicable, we will have to go back to the drawing board for other solutions. The institutional fabric of society must deal with the new destructive elements of global industrialization in a globalized economy.

A third line of reasoning brings climate governance deeper into the realm of political philosophy. One main purpose of constitutional arrangements for government is to safeguard citizens from usurpations of power, avoiding top-down decisions about what is good or bad for people (Aristotle, 330–322 BC). Conversely, absolutist philosophy such as arose in the 16th century (Bodin, 1576/1992); Hobbes, 1651/1968) gives central government this task, based on the rulers' wisdom or divine insight. This philosophy went against long-term developments restricting centralized power. With democratic procedures for representing the ideas and interests of citizens, absolutism became more acceptable again. In Institutionalism this process must be countered. Constitutions and broader institutions must primarily set the procedural rules for coordinating decentralized activities and for resolving conflicts. In addition, government should prevent private usurpation of economic power because it leads to political usurpation of power. Democratic welfare theory loses it prime position here in guiding public policies, including climate policy. The primary reason for having effective climate policy is to prevent disastrous instability in society caused by climate change, from the global to the national to the regional level of drowning cities. How citizens are to fill in their roles is in the hands of the citizens themselves. It is not a public task of government to guide their consumption toward zero emissions, but only a task to reach zero emissions for society at large. New rules are required for the main fabric of societies, their institutions and possibly also constitutions.

Institutions should certainly take notice of their possible welfare effects, but only as a second-place consideration. Ownership rules and the protection they offer form the backdrop for the still ongoing industrial revolution. Other more general mechanisms also have enabled Western societies to flourish, including reducing the powers of central government, opening up institutions for learning, and promoting decentralized organization. Through industrialization societies were able to develop and globalize. Further industrialization now requires new adaptations to the institutional framework, such as preventing the monopolization of unrestricted economic and political power. Effective climate policy is one important element for the long-term development of society, involving institutions that are yet to be developed for that purpose. Society must make a choice, either for absolutist action in a Planning & Control mode or for decentralized action through Institutionalism, the option detailed in this chapter.

Accommodating major trends

Instrumentation design must accommodate major societal developments surrounding climate policy and actively connect to such developments. These general developments constitute its regimes and landscapes, in design theory jargon. Climate policy has such a pervasive influence on society that it not only accommodates to such developments, but is a main constituting factor as well. The choice of governance in climate policy co-determines general developments in governance. Views on general developments in society ultimately depend on the more general ideas of the beholders. Strategic views in Institutionalist thinking cover social, economic and political developments and are different from those in a Planning & Control approach, as detailed in Chapter 4. Institutionalist views on main developments include the following.

Main trends are in connected domains. First, research and development (R&D) and also marketing have become irreversibly globalized, based on globalized information networks. There is tendency toward natural monopoly based on increasing returns to scale, where larger firms can produce items more cheaply. The firm that is largest reaps the most benefits in its domain, supported by IT and globalized markets. The tendency towards monopolization is strongly reinforced by the explosive increase in patent coverage, keeping out newcomers but now also increasing innovation costs. Conflict regulation with respect to contracts, intellectual property rights, financial markets and capital transactions represents an increasing share of GDP.

Next, primary income inequality within countries has generally been increasing over the past half century, creating political risks of populism and usurpation of power by the wealthiest. Today, global income inequality is decreasing, due to relatively high growth in newly developing countries, whose new middle classes outcompete the lower half of the workforce in developed countries. Inequality in wealth is increasing globally, and within nations is higher than income inequality. Political participation based on emancipatory action, such as that advocated by unions and ethnic and cultural organizations, has substantially collapsed in industrialized countries. Information circulated via the press, radio and TV, and increasingly the internet is primarily financed by advertising. Information supply is dominated by the richest 1% or less of the world population.

Third, regulatory processes tend toward centralization, to better address increasingly supra-local issues. They also tend toward detailing, to address broadened welfare considerations in an imperfect world. Regulatory processes tend to be dominated by specialists who are related to the private interests concerned. And these processes also tend toward internationalization, due to globalization in technology and production. For such reasons, political processes tend toward specifics, with a rise in issue politics and issue parties. There is broad political discontent in industrialized countries, with a majority seeing its share in the economy shrinking, its market position weakened, and its concerns not reflected in mostly technical or values and rights oriented political discussions.

The starting point for climate governance, broadly speaking, is whether to go along with current trends, or attempt to reverse then. The normative position taken in Institutionalism is that the main undesirable tendencies need to be reversed. The priorities to be set are the following:

- Centralization tendencies involving ever-more-detailed technical regulations must be reversed and replaced by generic incentives.
- Mirrored, detailed technical and behavioral regulations must be avoided as they are inevitably conducive to centralization and monopolization.
- Monopolistic tendencies must be reversed so that economic, social and political developments will remain open to broad participation.

It seems nearly impossible to reverse some tendencies.

- The winner-takes-all trend is deeply engrained in the new technologies related to automation, information technology (IT) and artificial intelligence (AI). However, this trend can be softened, such as by reducing the role of intellectual property rights in creating such near-monopolies.
- Inequality in primary income distribution can be reduced through better education for all and regional economic policy, but it will remain. It can only be softened, as by tax differentiation for more equal secondary income distribution. Public supply of services below cost, such as in health, education and housing, can create a more equal tertiary income distribution.

A clash between the two main governance directions is unavoidable, and this applies to climate policy instrumentation as well. But climate policy should not aim to solve all of society's problems at the same time – a herculean task that would be impossible to achieve within the relevant time horizon for climate policy of half a century or even less. Solving climate instability should be one prime focus for redesigning institutions.

Operational design rules for Institutionalism

The operational design rules for climate policy instrumentation discussed here are based on the instrument typology outlined in Chapter 3. Institutionalism starts at the most generic level, which is the level of institutions. It moves down to the more specific and binding instruments only as far as they are required to reach the 2-degree or 1.5-degree temperature rise goal; see figure 3.4 earlier. The design strategy is completely the opposite of the concrete regulatory approach of Planning & Control. That approach starts out at the bottom of the instrument classification and then moves up as far as is necessary to achieve the climate goal. The additional design criteria as related to administrative and economic efficiency are the same as for the Planning & Control strategy: simplicity, with sparseness, completeness, non-overlap and equality in instrumentation.

These main lines of Institutionalist design deviate fundamentally from current instrumentation, which is in a mixed, not yet coherent and effective, primarily Planning & Control approach. Institutionalism has faith in induced bottom-up development, replacing the control orientation of top-down planning. Bottom-up developments relate both to private actions and to decentralized public policies and actions, at both the city and regional levels. The uphill battle is replaced by a level playing field where ingenuity and initiative can be rewarded, including economic initiative, and a culture of emission reduction can evolve. When moving toward development of Institutionalist instrumentation, the starting point is current instrumentation, focused primarily on subsidies and standards and regulations, but with elements of a more generic nature. Major examples are emission pricing with cap-and-trade systems and emission tax systems that are present in several countries and states and provinces, but do not yet cover all CO_2 emissions equally and efficiently. Breaking with the current system must be feasible, and should occur as a normal stepwise change of direction, with many subsidies and regulations replaced and transformed into more generic incentives. Examples of instruments that would then become superfluous for climate reasons are fleet standards, insulation standards and other energy-efficiency policy measures covering many thousands of pages of public regulations in developed countries, mirrored by many more pages in the private firms that have to deal with these rules.

Internationally, as well, the differences with current practice and goals are deep. Generic instrumentation – that is, central and incentivizing instrumentation – leaves further policies and actions mostly to lower administrative levels and to private parties, enabling them from the bottom up. There is no reason to specify allowable emission volumes per country per year in international agreements, Planning & Control-style. A sanction mechanism is necessary, however, to take away advantages from those who do not implement climate policy. Some international cooperation on aligning instruments regarding emission-pricing is useful for that purpose, as well. Current distributional issues, with emission shares and transfer payments based on ethical principles, leave the domain of climate policy and its instrumentation. If there are good reasons to support developing countries, these could be part of such an independent policy domain. The main thrust for high growth in poor countries has been access to global trade, with hundreds of millions of people rising out of poverty in China alone. This development was not based on aid mechanisms, but on open rules for trade, substantially formed in the WTO, a largely limited-domain organization.[1]

Institutions repair and development

Institutional design addressing climate policy aims touches the heart of modern industrialized society. Ownership rules, liability rules and contract and trade rules, combined with the legal framework for conflict resolution, constitute the core economic institutions that engendered the industrial revolution as an ongoing process, supported by free knowledge creation in expanding schools and universities. In the background are issues regarding the independence of universities, freedom of information exchange and limited ingression of public bodies into private activities. Whatever is required to bring climate-changing emissions under control using adapted institutions has to fit in within this primary institutional framework. This developing framework has dealt with the specifics of environmental policy since the late 1960s by expanding liability rules to cover pollution problems. More generally, dealing with collective goods and natural monopolies have been issues of concern for many decades. The OECD's "polluter pays principle" aligns itself as much as possible with the principles behind liability rules: those who cause damages should pay for them. This broad principle has been interpreted in many ways, the minimum being that the polluter should at least pay for the cost of preventing emissions. Paying for the damages, too – the Pigouvian tax – is the more encompassing principle (Pigou, 1921/1932) but would require damage specification, which is hardly achievable in the climate domain.

For the purpose of climate policy, the more practical Baumol approach is followed: setting the tax at the level necessary to reduce this externality to the level deemed politically acceptable (Baumol, 1972), ofcourse depending also on other institutional instruments developed.

The electricity market was actively created and has continued to evolve since its start at the end of the 19th century. Electricity, with its production, transport and distribution all tending toward monopoly, was long supplied by public monopolies or extensively regulated private monopolies. However, long-distance transport over high-voltage transmission lines developed inter-monopoly cooperation, which subsequently invited competition. Privatization of production could then become an option, with large-scale producers competing against each other. Transmission and distribution grids remain strongly naturally monopolistic, however. Conservative neoliberal ideology has led to private ownership of these grids in most developed countries, with restrictions on ownership structure and operations. Transport systems for district heating are also naturally monopolistic, with some competition from electricity, heat pumps and solar heat. Pipelines tend to be privately owned, accompanied by ownership rules and behavioral rules to reduce excesses of monopolistic behavior. Diverse institutional arrangements have now developed, most of which accept private ownership of natural monopolistic infrastructure.

This is the context in which institutional instruments have to be developed, starting at liability rules and internalization of external effects and then moving to novel market designs for the emerging near–zero-emission economy to come.

Expanding liability rules for climate reasons

New liability rules could play an important role in developing climate policy. Such *strict, joint and several liability* rules were developed in response to toxic emissions in Japan and were used extensively to deal with toxic wastes in the United States and the Netherlands. If liability principles for climate damages could be fully developed and effectively implemented, all emitters would have to compensate all those suffering damages. The extended polluter pays principle would be implemented: the polluter would pay for the damages caused, and to avoid damages would pay for the cost of emission reduction. The core question in this regard is to what extent this can be made possible for climate policy at this most general institutional level. It is unclear whether it is possible to introduce new liability rules without also introducing more specific public regulations. This subject has been discussed in detail in Faure and Peeters (2011).[2] A first caveat respecting that discussion concerns the parties against whom courts can issue orders. One type of order is an injunction, or an issuance of injunctive relief, to force emitters into performing desired behavior, emission reduction, as a courts-based technology-specific regulatory instrument –an approach that is certainly not Institutionalist. An injunctive court order could also force governments to take action in order to reduce emissions. Forcing climate policy via court orders is also, at best, a piece-meal approach and is not a desirable option for efficient climate policy instrumentation. A court-ordered injunction to take certain action would then require some more specific climate policy instruments. Such court action may be part of a political nongovernmental organization (NGO) strategy but does not belong to strategic instrumentation design.

A converse relation is also possible. If other instruments have already been implemented, the application of liability rules becomes much more complicated. An emitter functioning under a cap-and-trade system has publicly acquired emission rights. Proof of wrongdoing is then difficult to establish. Similar reasoning holds when functioning under an emission standard. The European Environmental Liability Directive brings some order to this subject but does not include climate liability. The Directive has been implemented and has been the subject of legal actions.[3] For dangerous activities strict liability applies, which eases the proof of wrongdoing. For activities that are not directly dangerous, such as CO_2 emissions, fault liability applies, requiring proof of wrongdoing.

What requirements must be met in order to implement the full polluter pays principle via private liability law? The primary requirements are, first, proof of a causal relationship between specific acts and the damage and, second, culpability in the sense that the act should not have taken place, or that the harm should have been avoided. In Japanese, a case concerning the rise of sulfur dioxide emissions from 1970 onward seems to provide a good example of the application of the polluter pays principle through liability law. A causal link between emission and harm was established in this case, involving power station effluents and respiratory diseases. Damages were measured in terms of respiratory health costs incurred, and it was established that the monopolistic electricity producer was the nearly sole source of the damages. Class actions led to court orders to compensate the victims. The victims were represented by the government (often by local government), with total damages calculated in terms of health affected by the total of emissions, paid per unit of emission. Delivering compensation to the people who suffered the damages of respiratory disease proved too cumbersome. The liability system was subsequently transformed into an emission tax. Later that tax system was transformed into direct regulations (Broadbent, 2002). Significant payments based on liability have been realized under the Superfund regulations in the United States and similarly in Netherlands. The damages were measured in terms of the cleanup costs of toxic waste sites rather than in terms of harm to individual health, and were paid by the firms that originated the wastes to the cleanup agency: governments. The clean-up costs led to substantial payments by the firms involved. Payouts were the highest per capita in the Netherlands but covered cleanup costs of only some major polluters (O'Neill, 1990). Could such models be applied to climate-changing emissions, at least to those activities where it is possible to measure emissions?

Three basic problems need to be solved before liability litigation can be effective in the climate domain: establishing who caused the damage, the nature of the damage, and to whom compensation should be paid. The "who" causing the emissions may be one of many: the emitter, the final user of the fossil energy who is the economic cause of causing the emissions in the first place, the upstream primary producer of fossils, governments of current and former countries that failed to prevent the emissions, or the primary producer of fossil fuels. Since liability actions are linked to jurisdiction, holding a foreign emitter liable poses a number of very significant practical hurdles. If each emission anywhere leads to damages everywhere, all countries (and plaintiffs) could sue all foreign emitters. If governments are held responsible for failure to take action, whom should they pay? Themselves? This would make little sense, being merely a transfer from one pocket to another. All other governments? The final user is the ultimate economic cause of upstream emissions. But the electricity user did not ask for fossil production in particular: renewable and nuclear electricity are indistinguishable from the fossil-based electrons. So the actual direct emitter would be a prime choice as the party directly responsible. But then the first problem rears its head again: who are the emitters to approach, among the many billions of direct emitters across the globe – as private persons, as firms, as foundations and as governmental organizations.

The "damage" question is similarly double-edged. What is the damage, and related to the "who" question, which part of the damage is caused by whom? The principal problems in establishing the damage are the long lapse of time between emissions and the specific damages, which evolve over centuries; the conditional nature of damages, which depend on adaptations and include some positive effects; the global nature of damages; the highly nonlinear nature of damages; the high uncertainty of specific effects evolving, combined with other relevant mechanisms; and uncertain options for adaptation. These problems are similar to those that make the Pigouvian tax level almost impossible to specify. It seems quite impossible to attribute a share of partly known damages to a particular slice of emissions. Should only the damages that are caused nationally be accounted for, or a share in global damages? Emissions produced far in the past hardly contribute to the dangerous climate change now imminent, and current emissions can hardly be connected to future damage. If emissions were to stop right now, not just stabilize but decrease to zero, only a limited domain of damages would await us. It is the next cubic kilometers of CO_2, overspending a global carbon budget, that may set the most unacceptable changes in motion. Should only future emitters, then, be subjected to liability, per country? Reasoning points in that direction. But how should such future liability be designed, and would application of the new rules work? Could the liability payments expected in the future drive change now? Given the relatively short time span linked to economic decision making, this seems highly improbable.

Finally, to whom should compensation be paid? To the persons damaged? This would cover a significant part of the global population, including those who died prematurely as a result of rising temperatures, drought, flooding and hunger, and those yet to be born. Along the lines of a class action, a representative body would need to be appointed to receive and also spend proceeds on behalf of those damaged. Given that climate warming is a global problem, global government would be the ideal candidate. No such government exists in the sense of a governmental nation-state, and it is highly unlikely to emerge in the foreseeable future. Thus, national governments are the most likely candidates to receive and spend payments for future damages. For the dead and unborn this is an unworkable proposition, and it is hardly possible for the generations living now. And, of course, distribution among countries in accordance with reasonable principles would be very difficult to realize.

In sum, for very basic reasons private liability rules do not lend themselves to reducing climate-changing emissions systematically. Even if the questions relating to responsibility, causation, adaptation issues, and quantification of damages, including future damages and the possibility of extreme damages, could all be resolved, the system would still not function systematically. Purely private court actions are practically impossible to implement in many jurisdictions due to the large number of persons and organizations involved. The starting point for their design would be payment by emitters for their emissions, to be received by national governments. This looks very similar to pricing emissions with an emission tax, but has a very different legal and administrative footing. Internalizing external effects via emission pricing is a more direct administrative approach, with several highly feasible options.

Pricing instruments to create liability publicly: emission taxation

Internalizing external effects by a tax equal to the damage caused by an emission was first formulated as a policy instrument by Pigou (1920/1932), at that time with mostly local pollution in mind. Applying this economic tool to internalizing the external costs of greenhouse gas emissions would be extremely difficult, because of the impossibility of reaching agreement on the damage costs of an emission or of a certain amount of climate-forcing. Estimates such as those in the Stern report (Stern, 2008) differ significantly from other estimates and are seen as ridiculously high in the survey by Tol (2008). Yet even Stern's high estimate certainly misses many relevant effects, including the low-probability catastrophic effects described by Weitzman (2009, 2011). Even if inclusion of some catastrophic effects in integrated assessment models were possible, as was advocated by the U.S. Environmental Protection Agency (Kopits, Marten, & Wolverton, 2013), the principal effects of climate destabilization and social disruption could not be covered – assuming scientists could even agree on what then would happen, and what damages would follow. Further, the global damage costs of climate change, however established, will differ substantially among countries and interest groups. There may also be substantial agricultural benefits for countries such as Canada, Finland and Russia, leading to substantially different national Pigouvian tax levels for the globally occurring problem of climate change. This type of internalization of external effects per country has been investigated following a few attribution and equity principles. Extremely different outcomes result, several orders of magnitudes apart (Anthoff, Hepburn, & Tol, 2009). The Pigouvian tax is directly connected to these impossible-to-resolve quantification problems and hence cannot practicably be established for climate policy implementation.

What can be established is a price put on emissions expected to lead to a desired level of emission reduction (Baumol, 1972; Baumol & Oates, 1971). This requires empirical modeling of how society reacts to emission pricing and then, more specifically, how economic decisions are influenced toward achieving emission reduction. Although fraught with uncertainties and conditionalities, as well, this is researchable. Emission prices can be revised on the go on the basis of actual experience. If emissions should decrease much more rapidly than required to remain below a 1.5-degree temperature rise, or too slowly to remain below 2 degrees, the tax level can be adjusted. To avoid unnecessary costs during the transition, the price of emissions is set low when introducing the Baumol emission tax, but rises predictably to very high levels once emissions have been substantially reduced. In that way, the tax structure can steer long-term decisions related to public infrastructure, public and private R&D, and private investments. This pricing is very different from Pigouvian pricing, which is designed to be equal to marginal damages at any moment in time. With high marginal damages per unit of emission, the prices should start high. They can be lowered when the problem has been largely resolved, when marginal damages may then become lower. More extensive analysis of pricing options, also covering cap-and-trade systems, appears in Chapter 3. All emission pricing requires emission measurement. Measurement is possible for nearly all fossil CO_2 emissions, but less so for most other greenhouse gas emissions. Emission pricing is therefore focused on fossil CO_2, which is also the main cause of climate change. Equal and encompassing prices for all emissions is desirable for political-administrative reasons, for economic efficiency, and for ideas on justice. Reducing the number of regulatees is highly desirable as well, to reduce the public and private costs of regulation. The encompassing upstream-administered Baumol emission tax remains as the instrument of choice for internalizing climate effects in economic choices. Where cap-and-trade systems exist, as most clearly in the EU and California, these may be transformed in a stepwise fashion. First, step-by-step they move their administrative implementation upstream, ultimately to primary production and imports. In this way, their domain of application is broadened to cover all fossil emissions in all applications, while also covering all intermediate losses such as emissions from refineries. In parallel steps, the emission permit price is stabilized, with a floor and ceiling around mid-level. Their difference ultimately decreases to zero, making all trading superfluous. Finally, the single mid-level price moves along a predetermined path in time. The cap-and-trade system will then have evolved into an upstream-administered, encompassing predictably rising emission tax, with its legal status adapted accordingly.

Encompassing emission tax as carbon deposit

The CO_2 emission tax payment is implemented upstream at the level of primary production of fossils and their imports, and at imports of derived fossil energy products from refineries, as also proposed for the United States by Metcalf (2008). Derived fossil products produced within the tax jurisdiction have already been paid for by the tax on coal, oil and natural gas at primary production and imports. Downstream, a tax refund is paid upon export of such primary and derived energy products and upon approved carbon capture and storage (CCS), to be paid from the tax proceeds. All fossil CO_2-emissions are taxed this way as a carbon deposit (Huppes, 1993, p. 343ff) and (Huppes, 2011): there is no refund on what is lost on the way. Similar options, but with less clear system boundary definitions, were presented earlier in an OECD publication (Victor, 1992). Taxes and refunds at external borders are an administrative aspect of the national emission tax; they are *not* border tax adjustments. Border tax adjustments may be applied to products imported from countries without similar emission pricing, if substantial untaxed emissions are involved.

Boundary issues

Several boundary issues exist relating to what constitutes fossil fuels and fossil-derived energy products to be covered by an emission tax. Fossils would include peat for heat and electricity production, which would be treated according to carbon content, similar to light forms of lignite. Fossils would certainly not include stored straw and biomass fibers still composting in the field. But emissions from coal mining and natural gas production should be covered based on direct measurements there. A shift would be made to start taxing peat somewhere at a point of its aging towards lignite and then collection for energy use. The boundary may best be set where regrowth on the harvested biomass is possible; at that point it certainly is not fossil. Peat oxidation in peat areas used for agriculture, a considerable source of CO_2 from often centuries-old biomass, also seems beyond taxation. But options for measurement and then taxation could be developed, as soil management techniques relate to the amounts emitted. Burning peat to clear land would best be covered by other instruments. More complex issues arise where fossil energy sources are also functional products. Synthesis gas (hydrogen and carbon monoxide) produced nationally from coal would be covered at its source at coal mining. But where should imported synthesis gas for plastics production, the carbon in the carbon monoxide part, be taxed? A further boundary issue to be resolved is the heavy carbon content in non-energy products, such as in monomers for plastics production. And what about imported plastics themselves? Synthesis gas and bulk monomers could be included at import as part of the carbon tax, not being a border tax adjustment (BTA). Formed plastics might be taxed at import. Export refunding is due for products taxed when entering the emission tax system. On the outflow side, plastic applications exist which in the ground may last for thousands of years, turning into a sort of CCS. The amounts are small and the tax is usually not a significant part of costs; thus, taxation could open a Pandora's Box and seems unwise.

Basic reasoning from the perspective of simplicity would leave all downstream fossil-based products within the system and issue no refund on them. There may then be a reason for taxing products from non-taxing countries according to carbon content in order to create a level playing field. A rough cutoff criterion could be a minimum share of the tax relative to the value of the traded product. If more than (let us say) 30%, the product is part of the national system and the emission tax is applied. Then, a comparable refund must be given at export. For larger countries and countries collaborating in the tax system, these flows of carbon-containing products in the system will be very small compared with direct fossil energy flows. No credits or tax deductions at incineration of fossil-based plastics and other fossil-derived products should be given. If they were not taxed upon entry into the system, they could still be taxed then. That seems only marginally significant for emission reduction. Further, all biomass-based products remain outside the system: no credits at incineration. Their relative advantage lies in not being taxed for their biotic carbon content. The fossil carbon-based emissions from agricultural and horticultural production processes are fully covered by the tax, as in the price of the petrol and other fossils used there, resulting from the primary input tax.

Administrative implementation at the country level would be simpler than any other emission pricing system. The number of regulatees is a small fraction of the number of emitters, especially in larger countries and blocs. A working estimate is in the order of thousands. Tax evasion is always attractive. Due to relatively large-scale flows, very different from cigarettes and alcoholic beverages, measurement would be relatively easy and reliable. A certain amount of fraud will always remain, however. Implementation may take place at different administrative levels, corresponding to where other duties and excises are administered. This would be at the member state level in the EU, but may be more centralized in China and the United States. Refund at export fits into the same framework as for alcohol and cigarettes. Refund for CCS is based on certified long-term storage in well-managed and monitored locations, paid for by the government bodies controlling CCS. Individual countries may differ as to costs of storage, and it is attractive to reduce actual storage and refunding. No country now performs CCS regularly. No easy solution is available for that. However, the problem can at least be reduced if the tax is managed in larger blocks. First, the cheapest locations will be favored and the costs of CCS will somehow be shared by all participants from the total proceeds. It would then not matter where CCS takes place. Next, net proceeds of the emission tax, after export restitution and CCS, can be distributed according to share in actual emissions, regardless of where export or storage takes place. The actual total of emissions can be calculated using the tax and refund administration, but their location, and hence the distribution of proceeds, can only be based on modeling, through for example the use of input–output data.

An emission tax rises linearly, such as in the case of the British Columbia Emission Tax (Min-Fin-BC, 2014) the UK Price Floor Tax (Revenue & Customs, 2010/2012) and as intended for broader emission pricing in Canada. The UK Price Floor Tax would rise by approximately $5 per year (at £1≈$1.25), starting in 2012, leading to a CO_2 price in 2050 of around $250 per tonne. The Canadian price level would rise by $10 per year for the first 5 years. Combined with modeling outcomes, a provisional, modestly high estimate is $300 per tonne CO_2 by 2050, corresponding to a tax of $135 per barrel of oil, and similar for other fossils. The oil price for producers will be depressed in the long term due to reduced demand, dropping to

Table 5.1 Proceeds from a $200 emission tax in 2040

World energy supply 2040	200	TWh	After WEO2016*
Fossil share 30% (as natural gas)	60,000	TWh	Roughly as WEO2016
CO_2 from gas-fired power/heating	200	tCO_2/TWh	
Fossil emissions 2040	12	$TtCO_2$	
Emission tax	200	$/$tCO_2$	Assumed in 2040
CO_2 tax proceeds	2,400	T$	
Global GDP 2040	170,000	T$	Estimate WEO2016
CO_2 tax proceeds as % of GDP	1.4	%	
Total taxes as % of GDP	30	%	In between EU and United States
Proceeds CO_2 tax as % of total taxes	5	%	

Notes
GDP, gross domestic product; WEO2016, World Economic Outlook 2016; tCO_2, tonne CO_2; Wh, watt hour TWh, Tt, T$, tera ($10^{12}$)
*IEA (2016).

perhaps $30 per barrel of oil. The price for buyers would then rise to around $165 in 2050, close to the peak market price of 2008. Total net tax proceeds accrue to countries or to lower levels of government if involved. Countries may align themselves into taxing blocs, with taxes imposed and refunds issued only at outside borders of the bloc. Net proceeds are then distributed over these countries (or over states and provinces if so desired) according to their share in total emissions. This creates an incentive to not "forget" emissions, as that would mean forgoing proceeds. Using this approach, total net proceeds would first rise and then decline to low levels with low CO_2 emissions toward 2050; the tax and its proceeds would come to an end upon approaching near-zero emission society. However, some fossils production could remain, equivalent to the amount of CCS. As 100% effective CCS is not possible, this would require CCS on biotic carbon.

With this approach, the proceeds would first increase with the rising tax, and then diminish as a result of the induced emission reduction. A simple scenario for 2040 can provide an estimate of global tax proceeds as a share of global GDP and of public spending, assuming a reduction in emissions to only 25% relative to 2018; see Table 5.1. Only natural gas use is assumed at an emission tax level of $200 per tonne CO_2.

Instruments to create energy markets dominated by renewables

The core task in novel market design is to accommodate the substantial incoming share of intermittent renewables in most countries and regions, linked to an unprecedented level of secondary supply, both of a substantially decentralized nature. Intermittent supply can be matched to always-intermittent demand through several technical mechanisms.

- Compensating other flexible primary supplies, such as hydropower, and more flexible options for producing nuclear and concentrated solar thermal electricity.

- Compensating secondary production, such as by large-scale pumped hydroelectric energy storage (PHES), large-scale batteries and small-scale batteries and capacitors, and the active use of batteries, capacitors and hydrogen fuel cells, as used in transport vehicles.
- Reducing final use at times of mismatch between supply and demand, as with internet of things technologies.
- Spreading out peaks and troughs in supply and demand through long-distance electricity transport.

Most of these options do not fit into current electricity market designs. Most current markets have been unbundled to some extent but are still utilities dominated and do not charge price variation on wholesale markets through to most secondary producers and users, and certainly not on a real-time basis. The electricity markets will need to be transformed for this new challenge.

The first main institutional innovation required is the separation of markets at three levels, rather than the two levels most commonly found today. The level to be added is a high-voltage, direct current (HVDC) supergrid, operating on as large a spatial scale as possible, such as continental. Such a supergrid is currently being developed in and around China, for example. This supergrid can be expanded in virtually all directions and is potentially capable of linking to the silk roads in central Asia and next to Europe. The supergrid lines link regional high-voltage transmission grids, with alternating current (HVAC). These in turn link to lower voltage distribution grids, which are now mostly separately owned. In principle all levels are naturally monopolistic, being either publicly owned or publicly regulated. Producers and users are private entities. They connect to grids but do not own them. Public ownership can be decentralized according to the size of the grid. The supergrid will be largely national and international, certainly in fragmented continents such as Europe, and less so in large countries such as China, India and the United States. International connections are highly desirable in larger countries, as well, and have already developed to some extent, as seen in the linkages between Canada and the United States, and between China and some central and East Asian countries. The planned Desertec project (2003) to deliver solar energy from the Sahara to Europe would have involved such a long-distance HVDC transport grid, still expensive and technically uncertain at the time. How exactly such supergrid connections should be set up is open for discussion (Backhaus, Gausling, & Hildebrand, 2015; Rothe, 2016), both physically and in terms of ownership and market design.

Real-time pricing will develop as new technologies with microsecond reaction times evolve. The real-time price includes transmission costs paid to the grid operators, based on real-time congestion pricing (Verzijlbergh, De Vries, & Lukszo, 2014). Such congestion pricing avoids overloading of lines; it reveals where congestion is highest and where capacity increase is needed accordingly. Delivery to the transmission grid is open to any producer able to deliver at the required high-voltage level, including all large-scale producers, primary and secondary. The Nord Pool experience in northwestern Europe shows that supergrid systems connecting transmission grid systems can function even if they involve many partners.[4] Similar experience exists in the United States with interstate connections. Producers, primary or secondary, may choose not to deliver to the supergrid but to a more local transmission grid instead, and small low-voltage producers may link to the lower voltage distribution grids. These may then have to handle upstream electricity flows at times of high primary production regionally and low regional use. Transmission grids can also be international, such as the state-owned Netherlands transmission grid owner TenneT, which also owns a substantial part of the German transmission grids, owns and operates an HVDC line to Norway, and co-operates in several more HVDC lines, such as to the UK and Germany. Rules for operating as a public utility are difficult to establish in such a diffuse but substantially monopolistic system, however, and are avoided by separate ownership of the HVDC supergrid.

Intellectual property rights development

Current developments in intellectual property rights tend toward expanding domains of application, with increasingly detailed options and longer duration, expanding near-global coverage, and more options for enhancing protection of intellectual property rights. The option to reopen infringement litigation if a new argument can be made increases the power of large firms to exert their intellectual property rights, especially vis-à-vis individuals and small- and medium-sized enterprises. They can also press multiple legal actions to encroach on the rights of such smaller patent holders. This tendency began with James Watt, who defended his improvements to the steam engine using effective political action to extend his patent protection. Extended protection advocates presume that patents play an essential role in innovation. This was the main justification for patent protection until the end of the 20th century (Rohrabacher & Crilly, 1994) and is heartily supported by patent owners.

Views have started to shift, however. In the biomedical research domain, where the dominant view is that patents are essential to recover the cost of bringing products to market, research suggests otherwise (Heller & Eisenberg, 1998). They indicate that the recent proliferation of intellectual property rights in biomedical research actually leads to an underuse of scarce resources as a result of patent holders blocking each other. They refer to this as a tragedy of the "anticommons." This mirrors the tragedy of the commons, where lack of ownership leads to overuse of public resources, such as we see occurring in the climate domain. Shapiro focusses on a patent thicket, describing a dense web of overlapping intellectual property rights through which a company must grind its way to actually commercialize new technology (Shapiro, 2001). A number of partial solutions to this problem are possible. Patents can be dealt with in a cooperative way in common patent pools, in which the holders defend their rights as a group against outside infringements. The patents are shared in the group and the costs of upholding these patents against outsiders are shared by the members of the patent pool. This is an oligopolistic or near-monopolistic solution that is contrary to other decentralized initiatives. It does not serve a climate purpose in an instrumentation sense, but points to the density of patent protection as a problem that slows innovation. Another option under investigation is to increase the role of open, common standards. Patenting can be extremely attractive in technology domains, as patenting forces all others to pay licensing fees to the patent owner. In many instances it is possible to develop and apply open standards to the same domain, making patenting of such subjects nearly impossible. This may be a useful option in some climate-relevant technology domains. Examples would be the characteristics of high-voltage power transformation and the plugs for charging electric cars. The issuance of such standards belongs to the category of type 2 instruments.

Historical analysis also indicates that patents may have actually slowed technological development. After his first innovations, James Watt spent most of his time on legal procedures to prevent further innovations by others and stop them from entering the market. Only when his patents ultimately expired could innovations in the steam engine be introduced to allow for more efficient rail transport, mostly not supported by patents. The explosion in the number of patents granted since the 1980s, led by the United States, has given rise to high indirect costs (Jaffe & Lerner, 2004/2011; Moser, 2013). In fact, substantial historical evidence exists to show that *not* having a patent system yields clear innovation advantages, and so does not using existing patent systems (Moser, 2013). The modeling study by Joseph Stiglitz indicates that under reasonable assumptions, intellectual property rights may very well slow innovation by reducing the pool of knowledge available to all (Stiglitz, 2014).[5] Clear advice follows from (Boldrin & Levine, 2008). Their study is not historical-statistical, like Moser's, or based on general modeling, like Stiglitz's, but is micro-analytical and provides many historical examples. These are fully in line with Moser's statistical studies, and with the modeling approach of Stiglitz. The Dutch company Philips, operating in the Netherlands without patents, improved decisively on lamp design against the nearly global patents actively bought and defended by Edison, later General Electric. Their advice is to abandon any publicly supported monopoly, including temporary monopolies based on patents. The reasoning of Moser, Boldrin, Levine and Stiglitz is welfare related, in the liberal domain, and not related to libertarian freedom.

How much of this pertains to climate policy instrumentation? In energy supply and use chains, large-scale novel developments tend to be owned by a few large firms, with an increasing role for big data and AI, self-driving cars and smart grids being examples. Seeking patent protection is not a matter of choice for an individual firm. Without patent protection others can infringe on one's products by creating patents that overlap with non-protected technological innovation. Companies specialized in paralyzing patenting, known as patent trolls, already existed in the 19th century. Not taking part in the patent race means paying unknown but possibly extremely high fees to others. The consequence is that in the IT domain (Google, Apple, Microsoft, Facebook, Samsung and a few more), research costs are now dominated by patent costs and patent protection costs. Patent protection is widely used in the emission-free automotive sector, for example, by Tesla and Toyota, and in developing batteries for grid stabilization. Although abolishing the patent system altogether would be in line with Institutionalist decentralization and anti-monopoly lines of reasoning, the subject is so much broader that less-drastic changes in intellectual property rights should be considered for climate policy, certainly in the medium term.

One such medium-term option is to effectively reduce the domain of patent rights in most climate-relevant parts of new technologies. Developing countries have already demanded lowered patent tariffs for low-emission technologies, for example. This policy has moved to the forefront of international trade debates. Article 31 of the Agreement on Trade-Related Aspects of Intellectual Property Rights (TRIPS) allows national governments to issue compulsory licenses of foreign-owned patents in cases of national emergencies. This essentially lowers their price and changes the location of litigation, as well. For countries with different endowments, different regimes of patent protection for innovation may be reasoned, purely from an optimality perspective (Grossman & Lai, 2004). Compulsory licensing, as one of the instruments for helping developing countries, could serve more generally to reduce restrictions on the introduction of technologies for low emissions. This opening was seemingly intended under never-finalized WTO regulations, as stated in the Doha Declaration of 2001 (Article 5.b), that national governments have "the freedom to determine the grounds upon which such licenses are granted." Domain-specific restrictions on patent pricing could be a first option for climate policy. Domain specificity has severe drawbacks, however, as specific firms are involved, all of which would demand protection and induce chains of retaliation.

The first conclusion here on the possible role of intellectual property rights in Institutionalist climate policy instrumentation is that domain-specific instruments are not desirable because they increase complexity. Second, it is possible to conclude that reducing intellectual property protection in general is desirable and fits well into the overall reasoning of Institutionalism, including for preventing temporary monopolies. A third conclusion is that the reduction of intellectual property rights protection will take considerable time as it is difficult to reduce previously recognized legal rights.

Centralization, decentralization and separation of public and private

The core question with respect to the issue of centralization versus decentralization does not so much concern the levels and the tasks involved, as it does the type of regulations to be developed and, more generally, the types of governance required at different levels. There is an unavoidable tendency toward centralization when regulating specific private activities. It does not make sense for cities to set rules on the energy use of light bulbs, or the fuel use of cars. Going to the state or provincial level does not make sense either, as most markets are at least national, and most technologies are international. National smartphones do not exist. How to charge them has been regulated at the EU level for energy use and climate reasons, involving all major producers in the process. Extensive regulation has been set up in Europe via the Energy Efficiency Directive (with more than 1,300 pages of regulatory text), which forced all member states to implement the Directive's core rules in national legislation. It would be desirable for efficiency reasons for the United States, China and Japan to adhere to exactly the same energy efficiency rules, as firms develop products globally. Similarly, having global fleet standards for road vehicles would be desirable for efficiency reasons, if only for alignment of all test procedures. Such centralization may be justified by good welfare reasons, but it is to be avoided in Institutionalism, apart from some key standardization issues, instrument type 2.

Avoiding public–private partnerships

A compelling logic can be advanced for mixing public and private responsibilities when setting policy targets related to private activities. If regulation is clear and enforceable the results will also be clear. But if regulation is not clear and private parties can avoid compliance because it is difficult to specify rules and technologies are still developing, what must then be done? Markets must be corrected as compliance to reach targets does not evolve autonomously. If results can be closely connected to technology specifications, the subsidy instrument can be used. Where results are more diffusely connected to goals, the subsidy instrument is unsuitable and should not be applied. More interwoven contractual relations with mixed public–private responsibilities then enter the picture. The connection between firms earning part of their income from market operations and part from public provision, with risks distributed contractually, creates dependent firms and dependent government. As general rules for contracting are difficult to apply, the contractual relations are flexible and open. This approach requires specialized knowledge on the public side and is prone to complexity. Firms that are more often involved in public–private partnerships must also have specialized knowledge and adequate legal departments, and therefore tend to be larger firms. Specialized firms to arrange such contracts are also needed. Such a U.S. firm describes itself on its website to prospective clients as having the following specializations: "Connecting private and public sectors in a more efficient B2G marketplace, we create mutual value for companies, agencies and society at large."[6]

Public–private partnerships should not be confused with public procurement, even though their boundaries are not well-defined. Public procurement involves the acquisition of goods and services by governments from private parties. It is done according to special rules to control the setup of budget outlays involved and to avoid collusion and corruption. The reasoning behind such purchases is to create or support a public service, not to influence the operation of private parties. In public provision there is one private party (possibly a consortium) that receives the contract to deliver what is asked for. In public–private partnerships there is a competitive advantage for the contracting firm or consortium, very similar to a publicly established monopoly. One former public–private partnership was long dominant in the electricity domain, especially in the United States. The private firms involved in production and distribution had long-term contracts defining what to produce and how they would be paid for their activities, including their rate of profit. In that situation, the responsibilities of the parties, including for innovation, become blurred and innovation mostly stalled.

The core concern relating to centralized versus decentralized government is not primarily whether more centralization or more decentralization is best. What must be clear is the *type of task* that is to be handled centrally or decentrally. The first rule is to avoid technology-specific regulations of private activities as these lead to technocratic, centralized government, with blurred responsibilities. Private technology decisions are to remain with private parties, under possibly extremely centralized public institutions. Second, natural monopolies are to be operated publicly at the most appropriate level. Such public provision can be highly centralized if necessary, the most extreme case being that of a transcontinental supergrid that links national and regional transmission grids, but only for that limited domain. A continental supergrid requires continental public ownership embedded in encompassing electricity market rules. Distribution grids can be operated at the lowest administrative levels. Third, competitive activities should be private, while the development of private monopolies should be prevented. Fourth, public–private partnerships do not fit in with Institutionalist development. A mixture of public and private responsibilities should be strictly avoided regardless of the good climate motives that may be present. Fifth, substantial decentralized public tasks remain under Institutionalist climate policy design in addition to market design. Primary examples are in spatial planning, standardization, public transport, parking facilities, safety rules, educational systems, public research and more, generally in the repair of insufficiently functioning markets. Basic research hardly is commercially attractive and is public task. Another task is high-risk long-term technology development, which by necessity involves frequent failures and some successes, such as the internet developed by U.S. government and many medicines that were publicly developed, resulting in later commercialization.

Summarizing Institutionalist development of centralized institutions

- Options for applying liability rules in the climate domain are very limited.
- Pigouvian taxes are not feasible for practical and theoretical reasons.
- Internalizing external climate effects can be achieved by a Baumol emission tax on fossil CO_2, rising to the levels required for the 2-degree or 1.5-degree climate goal. This emission tax should be implemented at the national level and be internationally aligned.
- Energy markets should be developed with public provision of naturally monopolistic parts, and with real-time, equal, and open access to all private parties.
- Patent rights and procedures hinder innovation and should be reduced, preferably to zero, at the highest government level possible. This has relevance for climate policy reasons but has a much broader general advantages as well.

- Reducing patent rights is closely related to Institutionalist avoidance of publicly backed private monopolies, including temporary monopolies created by patents and adjoining copyright issues.
- Centralization versus decentralization of government is not the correct characterization of the issue. Most institutions are as centralized as possible, while allowing for a decentralized development of technologies and markets, and also infrastructure.
- Technological specifications and sector specifications in instruments are to be avoided as completely as possible, as are prescriptions on how lower levels of government should operate their specific activities or to reach specific targets.
- More generally, institutions in this model do *not* distinguish between sectors: liability rules apply equally to everyone, emission pricing does *not* differentiate between sources, electricity markets are open to any user and patent rules are to be equally reduced for all concerned.

Infrastructure, spatial plans and standardization for all sectors

Infrastructure here refers to public infrastructure only, not private buildings, for example. Spatial planning covers both public and private infrastructure and involves opening up some options and constraining others. Standardization primarily defines technical characteristics of products and installations and their connections, including services and measurement methods.

Public infrastructure

Public infrastructure is required only where natural monopolies tend to form. Management is aimed at preventing congestion in areas ranging from roads and railways to harbors and telephone lines. Natural monopolies may not last forever, although several have lasted a long time. What was once a natural public monopoly may later become one technology among many. Postal delivery services and fixed telephone lines were monopolistic for more than a century. Since the arrival of the internet, letters have largely been replaced by email and similar services. Fixed telephone lines were faced with competition first from cable services, and later from mobile phone connections. The wired internet (glass fiber and copper wires based ADSL) now receives competition from fast mobile phone connections, going from G3, to now dominant G4 to upcoming G5, reducing the monopolistic power of fixed connections Although all of these examples were once natural monopolies with public provision, this is now much less strictly the case. All such technologies have a tendency toward increasing returns to scale. However, easily leading to market dominance by single parties. Competitors can be acquired and small numbers of producers in a market may collude and set up rules to prevent newcomers from entering their market. Anti-monopoly policy instruments must then be developed and implemented, often boxing up against substantial economic interests. Keeping such infrastructure public reduces the risk of private appropriation of what effectively becomes a monopoly or near-monopoly. In weak political systems, the private monopoly may not be a natural monopoly but a publicly created private monopoly, such as can occur when giving the right of mobile phone provision to one private company. This happened in several countries, creating a few of the richest people in the world. Institutionalism certainly aims to prevent such a development. The patent system similarly creates monopolies, but for smaller domains of technologies. However, if many patents together cover a larger domain, the patent system helps to create private oligopolies and monopolies. When linked to climate-relevant infrastructure, such monopolistic tendencies can first be avoided by public provision, and next by adapting rules on the private operation of monopolies.

As a basic principle, infrastructure in the climate domain is supplied publicly where natural monopolies are present. Electricity grids are one main natural monopoly today and have been for quite some time. A distinction between HVDC supergrids, mostly HVAC transmission grids and lower tension distribution grids must be made. These three grid levels are owned publicly by a corresponding level of public body. In terms of infrastructure development, there are two main public tasks. One is to develop the large-scale HVDC network, preferably at an international and continental level, and beyond, as a supergrid. This will remain a main task for decades to come. East–west connections link time zones and north–south connections link climate and weather zones, easing the development of low-cost wind and low-cost solar systems. The second grid infrastructure task is to allow decentralized, primary and secondary production on at least the distribution grid, with upstream electricity flows. This is required when primary and secondary production in one part of a decentralized distribution grid or even transmission grid is larger than the total use there. Such technical adaptations seem part of the normal development of functioning grids and require some standardization of technical characteristics – a process that is already ongoing in many places today.

A few additional, possibly monopolistic, energy- and climate-related infrastructure examples are also noteworthy. One is warm water distribution system for heating in buildings, such as district heating. This is not a natural monopoly, however, as electricity with heat pumps may be a cheaper option in many cases already. Of course, it would be technically and commercially attractive to have public support by making district heating obligatory, for example, thus creating a monopoly. That monopoly could then be publicly owned or privately operated under public rules, as once was the case in many countries with electricity. There seem to be no good reasons to pick out such a technology as a subject for public infrastructure ownership, because the energy source for district heating is not necessarily zero emission. Where low-emission heat is cheaply available, local district heating may develop privately, as is the case with some geothermal sources and with waste incineration. There may be a spatial planning task involved, creating options for private development. This would be subject to normal spatial planning considerations, however, and not as a climate policy instrument.

Next, a hydrogen transmission/distribution grid could be developed publicly, linked to publicly owned hydrogen storage systems, primarily used in transport. However, in the short term this will have to compete with fossil fuel for transport, even if fossil fuels are burdened with the rising emission tax. The much larger electricity system will be competing in transport as well, except for long-distance transport. Currently, hydrogen storage and transport systems exist for industrial purposes, which are partly using pipelines already. Hydrogen fuel cell cars are on the market already, with distribution networks in development. It seems superfluous to make the hydrogen transport and storage systems part of climate policy instrumentation under Institutionalist circumstances. If pipeline transport and storage were to develop for hydrogen, spatial planning would have to accommodate them under normal decision-making procedures and with normal safety requirements.

Spatial planning

The spatial planning instrument allows focused development of public and private infrastructure, primarily covering land use designations and infrastructure for all modes of transport, including pipelines and cables. It may link to specific technologies. Where are large and small wind turbines allowed and where not, and where may hydropower be developed? Where are different types of mining allowed, and under what conditions? Most of spatial planning has some climate relevance. Climate-specific spatial plans can be set up as well, as elements in larger plans. Fossils may be phased out by reducing locations for primary fossils production, limiting pipelines and rail connections for their transport, limiting locations where fossil fuel vehicles may drive or park, and so on. Conversely, plans specify where renewable energy may be produced. The essential difference with Planning & Control spatial planning is that here options are created, but it is *not* a public task to fill in these options, unless to avoid creating private monopolies. When designating areas for large-scale solar PV, there is no public task to build them as there are enough private parties to do so, virtually all linked to the public grid.

Standardization

Standardization here refers to clarifying technology characteristics, including measurement methods. This is very different from binding behavioral standards for industry, such as emission standards and prescribed technologies. Such binding standards, instrument type 6, may be used for regulatory purposes, but not so in Institutionalism. National standardization bodies and the International Organization for Standardization (ISO) have standards to link products and technologies to their environmental and more specific climate performance. For the Life Cycle Assessment (LCA) of products there is ISO 14040:2006, 2016 and for Environmental Technology Verification there is ISO 14034:2016. They are widely used privately for commercial and also for ethical reasons. These standards operate privately but governments play a key role in selecting subjects and helping to formulate them. Avoiding provider-specific monopolization is a main goal in Institutionalism. For example, car brands should not have proprietary plugs for charging only their brand of electric car; instead, all providers and users should use the same agreed-upon standard as now exist for gasoline, either privately agreed or publicly enforced when private standardization is insufficient.

Research and development for innovation for all sectors

The current trend toward publicly funded research with valuation potential should be reversed. Public funding should be primarily for basic research in domains where private parties cannot expect commercial applications, or where private supply would lead to undue monopolistic development. This also holds for data and modeling of long-term climate development, where commercial exploitation is virtually impossible. Weather modeling and prediction is an example of a research activity where private monopolistic exploitation should be avoided. Funding should support research into long-term, possibly interesting subjects; in some cases there will be also climate reasons for setting up programs. Basic research without direct applications forms the core of public funding, through universities and research organizations. Some research may be seen in the light of potential applications. But certainly the purely precompetitive, publicly funded research must be openly available in order to contribute to further scientific development and must remain open for commercial exploitation by whomever sees commercial potential, creating the publicly available knowledge domain (Acemoglu, Johnson, & Robinson, 2005; Mokyr, 2004).

Domains of funding may be envisioned at an aggregate level. However, determining which research is interesting, and which is not, should not be decided publicly in a centralized decision procedure. Some randomness may be useful, to allow for out-of-the-box subjects. Playful funding mechanisms can be developed, such as in the DARPA projects in the United States and in contests held around a specific theme. The Kremer prizes for human-powered aircraft, for example, were funded privately and have played a clear role in developing human-powered aviation. Human-powered aircraft will likely never be a private commercial endeavor. Nonetheless, its successful development, including crossing the English Channel and flying nearly200 km from Crete to the island of Santorini, has been fun for many students. it has also refined the basic technology required for unmanned solar aviation with commercial potential, which might for example be used to bring internet connectivity to poor and sparsely populated regions. This is not a part of planning but of the option creation in a non-commercial domain.

Applied research, as in R&D, may be placed on the public funding agenda but only in a very restricted way: where commercial development seems important and cannot be expected to evolve toward commercial application due to high costs, high uncertainty and long duration to gestation. An example is large-scale nuclear fusion energy. Purely private development is absolutely not viable in this case. The number of such large special projects may be small, however, as complex systems usually are built from smaller ones. They can be funded either in generic surroundings such as universities, or can be developed privately for specific shorter term commercial applications. The current funding of new, high-energy-density battery systems may shed light on the boundaries to be considered. Currently, research in novel areas is predominantly publicly funded, with results going into private start-ups derived from university work, the university being commercially interested as (co-)owner of the patents involved. From patent income the university can then fund new research with commercial potential. In the Institutionalist line of reasoning such systems should however be discontinued. Basic research is publicly funded at universities, without patenting and licensing. Private development may occur in larger private research institutes or in smaller firms funded by investment firms focusing on high-risk/high-profit options, using knowledge which has been made publicly available. Universities must no longer function as market-oriented, semi-private organizations. Market-oriented private research then will no longer be chased by unduly publicly funded competitors.

In Institutionalism, one prime direction for public R&D is instances where technology is likely to become a natural monopoly that would be publicly supplied when developed. The main well-studied example of this is the fully public supply of the internet in its first years, resulting from international work at CERN in Geneva, where scientists worked on a system for easy communication and data sharing, and the U.S. Army, which worked on information exchange systems that were invulnerable to destructive local attacks. The U.S. Department of Defense's Advanced Research Projects Agency (DARPA) project resulted in the military development of the experimental ARPANET. An exchange between mainframe computers and scientists at universities followed, and this was eventually opened up to many private users, including telephone companies, and then to additional, highly regulated, private development of new parts of the system. The privatization of main parts of the physical infrastructure is ongoing, leaving monopolistic elements to cooperatives of private parties as quasi-public organizations, and with ongoing discussions on the net-neutrality of the system. The substantially monopolistic nature of the internet would be a reason to avoid privatization before effective competition can be organized. The semi-private ownership of parts, however, also provides a shield from public monopoly misuse. Publicly supplied monopolistic infrastructure does not necessarily require public R&D. The development, building and maintenance of dikes and long-distance electricity grids can be fully private, yet paid for publicly, and fully publicly owned.

There may be other reasons to endorse limited, publicly funded RD&D with actively supported implementation for creating learning curves, to overcome lock-ins against already developed privately supplied, high-emitting technologies. Solar PV, wind turbines, high-energy density batteries and electric cars are examples. Without a doubt, subsidizing solar and wind power generation has created learning curves that lowered their investment costs dramatically. But in an Institutionalist governance context, such RD&D should, for the most part, not be a public task. A generic emission tax would make many such developments interesting to private investors. For example, the use of the lithium-ion car battery in wearable electronics and laptops was developed mostly through private investment. Finding an example of a definitely relevant public development project or program that can be further developed as a market activity does not seem easy. There may be an essential public task in infrastructure development and standardization, to prevent private monopolistic tendencies. Standardization of connecting cables and plugs for charging and discharging electric car batteries, mentioned earlier, is an example. Supplying the standardized electricity supply systems is not. This option for publicly funded RD&D and public development is more limited than some advocates would prefer (Mazzucato, 2013); they would more generally include any major commercial options not yet developed by private parties, possibly competitively developed in public–private partnerships. In institutionalism a clear boundary between public and private is maintained.

Regulatory instruments for specific sectors and technologies

After developing instruments at the institutions level (type 1), the infrastructure level (type 2) and the level of R&D (type 3), it must be determined what instrument tasks remain for more specific technologies and behaviors in the private sector (types 4, 5, and 6). There are two main reasons to engage in regulating private actors: emitting activities have not yet been covered by generic instruments, and specific circumstances prevent higher level instruments from functioning effectively in particular deep market imperfections. One element of market imperfection is the restricted time horizon for innovative development due to the high risks accompanying deep innovations. Such innovations may then not be undertaken. This market imperfection is covered by publicly funded R&D, type 3. High risk investment also pertains to new and as yet unknown infrastructure. The internet would not have come about privately (Mazzucato, 2013). Investment in hydrogen infrastructure may be essential for a zero-emission society, or may be mostly a waste of money as renewable electricity with storage systems may fully outcompete hydrogen systems in a real-time electricity market. Because this cannot be known in advance, a public task may lie in developing the hydrogen infrastructure to a competitive level, as is coming up in many countries now. Government takeover of responsibility in a domain increases the risk of exclusively private market development since those receiving the subsidy can outcompete the purely private developers, even if the private sector designs are better. This crowding-out mechanism can be reduced by more generic risk-sharing, and by subsidizing developments with a clear public good character, such as lower emission options (Martín-Barrera, Zamora-Ramírez, & González-González, 2017). Such detailed technology-specific regulations don't belong to the Institutionalist repertoire. Making the low-emission option more attractive by taxing the competing emitting technologies is the preferred mechanism.

Thus, the basic question when considering technology-specific public policy instruments is whether they are really necessary under well-developed Institutionalist conditions, assuming relevant infrastructure implementation and publicly funded R&D for long-term technology development. The fact that subsidies may work on a case level is not a sufficient argument for the use of subsidies. They will to some extent ruin broader, normal market development, which is overwhelmingly important in industrialized countries, and the technologies chosen by governments may well induce development of ultimate loser technologies rather than winners. What, then, is really required in the four main sectors of the economy in terms of public regulations and public subsidies regarding specific private technologies? And where might such a public planning approach be essential for large-scale private introduction of low-emission technologies, given the encompassing emission tax?

In the energy sector, deep transformations will also be induced by real-time electricity markets and underground heat and cold storage ownership; well-developed supergrids, transmission grids and distribution grids; and a rising upstream administered emission tax. Are additional technology-specific instruments needed? It would be better not to develop smart grids through public regulation, as this will quite certainly be at variance with real-time electricity markets; currently publicly prescribed meters, for example, are not real-time based but instead are time–period based. Large-scale solar, wind and tidal power will require infrastructure and spatial planning, which are not technology-specific instruments. Should nuclear power be subsidized, apart from deep technology innovation? No. Its competitiveness would derive from emission taxes, while abandoning all explicit and implicit subsidies it now receives. The conclusion here is that no technology-specific instruments are required in the energy sector; existing ones can be scrapped, including all sorts of efficiency regulations and subsidies.

In the transport sector, the main bottlenecks toward achieving zero-emission transport lie not in personal transport and light freight transport, given that alternatives exist and will become more competitive under generic public task instruments. The main issues have to do with long lifespan fossil fuel–based capital goods, such as those used in heavy road transport, long-distance shipping and aviation. With their sunk costs, they can compete on only variable costs, while new low-emission vehicles are also to cover the capital costs in their prices. Public R&D will be required, probably very substantial in the case of zero-emission aviation. But a substantial economic advantage of the new non-emitting technologies could well drive out the old ones regardless, just as kerosene turbines drove out gasoline-based propeller aircraft on a massive-scale in the late 1950s. If the transition is not fast enough, a scrapping subsidy or emission standard can be used to drive out the old – an option that should only be considered several decades down the line. In aviation, current subsidies realized through tax reductions and exemptions based on international agreements, such as with the ICAO (International Civil Aviation Organization) should be abandoned, with the effect of somewhat reducing the extreme volume rise of aviation. However, this would not be an aim of Institutionalist climate policy: emission reduction is to be realized with generic means.

In the buildings and appliances sector, heating will have to abandon fossil fuels, the main source of emissions in that sector. Ground source heat pumps are one main option and can also be used for efficient cooling. Under generic

Can hydrogen-based aviation become competitive?

Some back-of-the-envelope calculations may give an indication. With oil prices at well around \$40 per barrel and an emission tax of \$300 per tonne CO_2, the price per barrel of oil would rise to \$175. Assuming linearities, the price of aviation fuel then would rise more than fourfold from around \$0.40 per liter to around \$1.75 per liter, equivalent to \$0.175 per kWh. The much higher energy density of hydrogen, by a factor 3, could give hydrogen an advantage, especially on long-haul flights, if additional storage mass would not outweigh this advantage. This is currently the subject of public and also private research (Bicer & Dincer, 2017; Xu, Li, & Huang, 2015). Bicer and Dincer see hydrogen fuel costs as roughly twice current fuel costs for medium-distance flight. The orders of magnitude seem to fall within a quite feasible range for hydrogen flight. Fuel cell-based electric drives could have a higher energy efficiency, with low-noise electric propellers for slower, short to medium-distance transport. For very short distances, battery systems may also be relevant, with small-plane experiments now ongoing which are mostly publicly (co-) funded. A survey of substantial technical developments is in the 2018 AIAA/IEEE Electric Aircraft Technologies Symposium, see https://arc.aiaa.org/doi/book/10.2514/MEATS18.

Institutionalist conditions, a private market opportunity is created. The price differential between electricity and natural gas resulting from the emission tax is key, next to a variable-priced electricity market. This differential could be created by other means as well, such as by subsidies. However, such measures have structural disadvantages, due to funding costs and their technology-binding nature. They cannot improve on the tax. For older buildings not suited to low-temperature heating systems, the cost of a changeover to non-fossil heating may be prohibitive unless combined with significant refurbishment for other reasons. Such refurbishment takes usually place after two or three decades, which means a major portion of all buildings will be refurbished before 2050. Instruments such as the Energy Efficiency Directive (EU, 2012), with more detailed implementation plans in 2016,[7] require national governments in the EU to refurbish a certain percentage of dwellings per year, with regular controls. Such instruments are completely superfluous and damaging for normal incentive-driven innovations. Induced refurbishing may not be effective in practice and may well have broad disadvantages from a life-cycle perspective, different in different regions (Crawford, Bartak, Stephan, & Jensen, 2016; Rosselló-Batle, Ribas, Moià-Pol, & Martínez-Moll, 2015). Leaving to the users and owners such private decisions about shifting away from expensive fossil fuels is clearly the preference of Institutionalism. The alternative, regulating tens of thousands of possible products, seems a Sisyphean task. Even substantial success in specific cases may not provide sufficient justification for large scale detailed measures. Incandescent light bulbs were phased out in the EU and replaced with fluorescent energy-saving lamps; however, LED lighting came soon and is substantially cheaper and more versatile, making regulations superfluous. Finally, in the buildings domain no technology-specific policy instruments are required, and existing ones can be scrapped for the most part. Overcoming deep lock-ins for appliances could, incidentally, be useful if they indeed produce significant emissions. This is likely almost never the case.

In the industry sector, emissions have a most diverse background, ranging from natural gas- and coal-fired heating in refineries and buildings, to the use of carbon in chemical processes, such as for reduction of iron ore in iron and steel production, engrained in historically developed cola-based technologies. Technically, a shift to hydrogen- and electricity-based systems is very well possible. Whether it is economically attractive in Institutionalist climate policy depends *inter alia* on the level of the emission tax, and on new process designs. Even at modest tax levels, hydrogen technologies may become competitive for new installations. The sunk cost of existing installations will require a very substantial cost advantage to overcome this lock-in, or patience and perhaps a scrapping subsidy much later. No further industry-specific climate policy instruments seem required.

Institutionalist instrumentation from global to local

A centralized framework for decentralized action

The public bodies involved in climate policy range from fully global to completely local, and all government levels in between. The private parties involved also range from the local to the global, and include citizens, consumers, NGOs and cooperative organizations, small firms and multinationals. Who is to do what and when? The answers very much depend on the style of governance, which is mostly determined at a national level. In Institutionalism the supranational level is required but has only a limited scope. There are no global emission targets, no country targets and no climate-based redistribution mechanisms. There are international agreements between main countries on setup and level of a generic emission tax, but hardly any such fully global agreement. The main climate instrumentation is aimed at the level of the nation-state. The main thrust is to incentivize and to facilitate decentralized actions. Diversely mixed decentralization options seem to exist side by side, such as applying different emission pricing systems in different provinces, Canada being a current example. Although possibly an expression of political reality, this choice is certainly not derived from general Institutionalist reasoning and also does not conform with some of the general simplicity design rules of sparseness, completeness, non-overlap and equality of application. More instruments than strictly necessary are used, with substantial and also complex administrative boundary issues, and emissions are not treated equally in terms of pricing level. Institutions are to be developed at the highest administrative level, including international alignment. All more practical decisions ensue at lower administrative levels, and privately. Spatial planning and local infrastructure development fall largely in the domain of local governments and in more detail in the hands of private parties, giving them real independence instead of distributed tasks (Huppes & Huele, 2018).

The nature of climate policy instrumentation is not the only factor affecting local independence and incentivizing decentralized creativity. But the choice of governance can certainly have an influence on local autonomy and the space available for bottom-up action (Capano, Rayner, & Zito, 2012; Ostrom, 2010; Rayner, 2010; Schroeder, Burch, & Rayner, 2013; Huppes & Huele, 2018). Bottom-up action is completely opposite from what the EU Energy Efficiency Directive and Ecodesign Directive stimulate; for example; they specify the rate of yearly building refurbishments required per country, then countries obligating cities to install programs to reach those targets. Under institutionalist governance such refurbishment might be organized voluntarily, also by cities, under changed pricing conditions, if this is considered advantageous from their integrated perspective, including climate considerations. But why should cities be made responsible? The city role may be very limited if private firms and organizations, including building societies, are already organizing larger scale refurbishment. Such decisions do not take place at the central level and can be made very differently per city according to local circumstances.

International cooperation beyond Paris

The United Nations–brokered Paris Agreement (2015) sets pledges for emission reduction per country. In the Planning & Control approach, the task is next to make the pledges more binding and the reductions deeper, in order to bring them in line with the 2-degree, or preferably 1.5–degree, climate stabilization goal. The Institutionalist international strategy, in contrast, involves equal design and establishment of the level of the rising emission tax and opening up options for regional international trade in electricity, with adjoining measures to reduce undue competitive disadvantages of non–climate policy countries. Because of its greater political and administrative feasibility, some experts prefer this type of agreement over binding caps (MacKay, Cramton, Ockenfels, & Stoft, 2015). But the basic choice in all versions of Institutionalism has a much deeper background than mere practical feasibility. The society envisioned is not a planning-oriented one, but a decentralized one, with institutional boundaries rather than technical boundaries. Although preferably as multinational as possible, the agreements may well start between two main countries, open for others to join under the same conditions. The original setup of the WTO is an example of an international agreement of a limited nature; the Hanseatic League in the late Middle Ages was another. The internationally aligned climate policy instrumentation of modern Institutionalism fits within this pattern of limited domain agreements. Alignment is seen in the setup of the national emission tax as an upstream-administered, excise-type of tax, and next, in the level of the tax, which increases over time. That level depends on a global view of the level needed to reduce global emissions to meet the 2-degree or 1.5-degree temperature-rise goal. These agreements are followed by practical arrangements on administrative cooperation, for reasons of efficiency and to reduce border controls. When countries join to act as one administrative unit for such climate policy, the net proceeds are distributed over the participating countries according to their shares in total emissions. The required measurement system can be comparable to what has been developed in the United Nations Framework Convention on Climate Change (UNFCCC). A higher share in emissions means a higher share in the net proceeds, which should lead to relatively honest reporting. Under-reporting on emissions will lead to a lower share in a country's proceeds from the tax. Over-reporting the level of emissions will induce the partner countries to check numbers and correct them.

With emission tax systems established, differences between countries will remain, with other countries not taxing or imposing only a small tax. The national emission tax system implemented upstream covers the actual non-negligible carbon content in imports, as part of the national emission tax system. It does not cover upstream emissions abroad, leading to competitive disadvantages for the climate policy country. Iron, steel, fertilizer and organic base chemicals are primary examples. Border tax adjustments are then justified in principle and are allowable under WTO rules (Tamiotti, 2011), with complexities in establishing a clear basis for such import taxes and symmetrically export subsidies and their effects on international trade (Droege, Asselt, Das, & Mehling, 2016). The UNFCCC uses the term *border carbon adjustments*, confusingly mixing upstream-administered national emission taxes with taxation of further upstream emissions that occur abroad (UNFCCC, 2016). These are two very different categories: out of them only the second one is a part of international trade law. As a first step, the tax level to apply per unit of emission abroad may be that of the national tax level. Determining how much has been emitted is less clear. Aluminum may have been produced by geothermal or hydropower, and this particular batch from this particular factory. But if homogeneous materials are traded and come from a warehouse, specification of origin may become difficult, and quite useless from a climate regulatory point of view. Furthermore, national competitors having to pay the emission tax will adjust their emissions over the course of time, ultimately to zero. Their production costs after adjustment are lower than those of the foreign producer who is paying for emissions that are still high. The tendency to overcharge foreign competitors may lead to trade conflicts and contribute to trade wars. One incentive induced by clear criteria for border tax adjustments is that the exporting country levy the emission tax itself, thus at least not forgoing the tax proceeds (Dong, Ishikawa, & Hagiwara, 2015). A partial emission tax system could then evolve there, possibly as a first step toward joining the emission tax club.

International electricity markets

International electricity markets are defined in terms of the HVDC supergrid. Their development, connecting all major transmission grids over regions and countries in a separate organization, is so advantageous in terms of voltage stability and cost reduction that for private economic reasons arrangements will be sought. The international agreements required relate to mutual access to national markets in national transmission grids and to some limited technical standardization. HVDC supergrids can connect very diverse transmission grids, with different frequencies and voltages.

International CCS agreement

For CCS, different international agreements may be useful to open options for using the most attractive and low-cost locations for CO_2 storage. With a refund paid from national proceeds for national CCS, the amount of CCS depends on the capture, transport and storage cost per country. Exporting CO_2 to another country with cheaper storage options becomes attractive, but not for the importing country that is paying for the storage. Including this flow in the export refund without further controls would very likely lead to massive fraud; it must enter another country and be taxed there, with a tax refund following upon control of the storage. CCS for 10% of current emissions, four billion tonnes of CO_2, taxed and refunded at $250 per tonne, would involve a payment of one trillion dollars by those providing the carbon storage. Before entering into international deals of this magnitude, clear administrative and technical procedures are required. Incentives move in the wrong direction if a "leak" can avoid real costs and collect refund money.

Summary of international climate agreements

The main ways in which Institutionalism differs from Planning & Control with respect to international agreements can be summarized as follows:

- There are no agreed national emission targets per year; instead, a global reference volume suffices for setting the emission price path.
- There is no universal global agreement, but a first step is taken with core countries, with others open to join this "climate club" on the same conditions (Brenton, 2013; Falkner, 2015; Falkner, Stephan, & Vogler, 2010; Nordhaus, 2015).
- There are no lower tariffs on "environmental goods," as was intended in the Environmental Goods Agreement in the Doha Round of the WTO.
- The deposit-type emission tax incentivizes exporting countries to join the climate club, avoiding the emission tax of the importing country by levying the same tax on its exports itself (Dong et al., 2015).
- A BTA is levied on high fossil energy products only, covering their undue cost advantages in terms of not being charged by a similar national emission tax. This is easily in line with WTO rules (Tamiotti, 2011), but difficult to specify empirically.

Transition toward Institutionalist instrumentation

Anticipated bottlenecks in this transition relate to a number of highly diverse mechanisms and effects. The first concerns replacing existing instrumentation by Institutionalist instrumentation without causing damage. Next, unintended side effects of an unacceptable level may need to be resolved. As climate policy deeply affects energy supply and use, old industries will die and new industries arise. Energy supply is affected in the process, with a possible bottleneck around supply security.

Development of institutions

The development of institutions involves electricity markets, heat and cold storage ownership and the emission tax system, each developing from its own current national status and options. An operant cap-and-trade system can, for example, be stepwise transformed to shift from measurement at the source to upstream measurement at the points of import and primary production. This enables all fossil fuel emitters to be covered equally. Next the price must be stabilized by adjusting permit volumes to the stated intended price. This effectively transforms the auction price into a tax, with administrative adjustments following accordingly. Starting with a low emission tax can avoid a possible shock of introduction.

Replacing existing instrumentation

Existing instruments differ among countries, and even within countries. Cap-and-trade systems can be transformed into an encompassing emission tax, see above. Subsidies contracts given cannot be revoked. Feed-in tariffs and premiums have a fixed end date. Such schemes may remain operant until they expire or become irrelevant. In general, all subsidy schemes can be reduced with the rise of the emission tax or may just slowly be phased out to avoid the costs of bankruptcies. Tendering procedures for wind power complexes can first remain roughly as they are but will become superfluous. Fleet standards will become irrelevant and can then be abolished, and similarly so for CO_2 emission standards.

Resolving extreme income effects of effective climate policy

A slow buildup of the emission tax will also generate a slow buildup of effects on income. There are no intrinsic reasons why Institutionalist climate policy would have especially negative income effects as compared with other more costly effective policies relative to the current situation. The emission tax proceeds can first be used to reduce other taxes, the revenue-neutral option as has been introduced in various provinces of Canada (Min-Fin-BC, 2014). The revenue neutral option also allows for redistribution, reducing regressive taxes. An alternative is to spend proceeds on new public outlays, favoring lower income groups most affected by the price effects of Institutionalist climate policy instrumentation. Or, the proceeds can be used directly as an income subsidy, paid out to households on an equal footing. This is a policy proposal introduced by Republicans in the United States (Baker et al., 2017). Finally, the tax proceeds might be used to subsidize renewables and clean, low-energy investments. Such subsidy instruments clearly do not belong to the Institutionalist repertoire. Are there other options for spending? The essence of Institutionalism is to repair what is deficient or to add what is desired at the institutional level. If unequal income distribution is an issue, whether for reasons of justice, economic growth or political reasons, that issue is to be traced back to income policy on its own merits, ranging from schooling to labor market design to regional economic policy, including industrial policy. Similarly, redistribution, such as that provided through progressive taxes, income subsidies and public outlays, can, but should not, be linked to climate policy. Other institutional arrangements might better be chosen, such as those related to the provision of effective education, the structure of the tax system and the supply of public goods. The increased income inequality that has developed in several industrialized countries over the past few decades (Milanovic, 2016; Pickett & Wilkinson, 2015; Piketty, 2014) is far beyond what climate policies might induce (Alvaredo, Chancel, Piketty, Saez, & Zucman, 2017).

How can income effects be specified, relative to which alternative situation? The before-and-after comparison is one way, assuming fully "Paris-effective" climate policy to come. Energy costs may very well increase and have to be paid for somehow. (The "no further" climate policy option would be akin to the with-and-without comparison often used in cost-benefit analysis.) But would lower income groups be most affected in this before-and-after comparison? The induced phase-out of fossils would depress their prices long term, while the rising tax, a transfer payment, would make the shift to renewables attractive. Overall real cost would rise, but not necessarily substantially, because cost saving options would also develop. How the tax proceeds are used, for which tax reductions or additional outlays, is not a matter of climate policy. A second option is to compare both potentially effective alternatives, Planning & Control and Institutionalism, in reference to the same future period of time. Real costs would be higher in Planning & Control, and would have to be paid for somehow, mostly in terms of rising prices of goods and services using the higher-priced energy. Subsidies, progressive or regressive, would be widely used, also in the form of tax deductions. Some taxes would have to be raised to fund the subsidies. These, again, can be regressive and increase income inequality, or may work the other way around. The emission tax in Institutionalism will work out somewhat regressively, with an option to correct for this effect in using the tax proceeds. The overall effects on income distribution could be larger in Planning & Control than in Institutionalism, but again, there is no strict reason for this to be the case.

One further element that differentiates the instrumentations is the open-to-all, real-time, variable-priced electricity market, which is present in Institutionalism only. This creates the option to use electricity, preferably at lower-priced moments. The price difference can be substantial, up to several factors, for example when charging ground source heat storage at low prices and using car batteries for power load equalization, even creating income. This last option may be a considerable advantage for lower income groups in developed countries. In lower income countries, with more limited car ownership, this option may affect only the more affluent segments of the population, those who can afford a car. But poor households there would not pay for the relatively expensive transport electricity.

Energy supply security during transition to zero emission

Energy supply security has two very different loci of possible disruption. One is dependency on foreign supply by a near-monopolist supplier, such as OPEC. Closing off supply can be used not only for economic reasons, to raise the price, but also for political reasons. Russia has interrupted natural gas deliveries for such reasons, and China has disrupted neodymium deliveries necessary for wind turbines and electric cars. This problem is deemed especially relevant for fossil energy; however, this source of supply insecurity will be reduced to near zero in the next half century, resolving that issue. Electricity can also be delivered from low-cost production countries abroad, such as hotter and drier regions for solar electricity, windy areas for wind power, and high lakes for stored hydropower. Given the low cost of long-distance transport using the public supergrid, and a large variety of sources connected to this grid, this first point of disruption affecting energy security hardly seems an issue in Institutionalist climate policy instrumentation. Of course, the United States should not rely exclusively on Mexican and Canadian electricity, nor should the EU rely solely on North African electricity.

The second locus of disruption is more internal, as in not being able to balance the grid, resulting in local to regional interruptions. Such brownouts are increasingly disruptive in a more electrified society. Again, large-scale electricity markets with interconnections among transmission operators are one solution; another lies in real-time pricing, which is more subtle. With the development of the internet of things, owners can program their appliances, including car batteries, to directly respond to price changes not as an administrative step requiring an interval of time, but in a real-time automated manner, which directly helps balance the grid. Such options are more limited in the planning-type, top-down version of smart grids using time-of-use rates. Overall, effective climate policy will reduce the risk of supply disruption, more so in the Institutionalist version than in a Planning & Control approach.

Institutionalist instrumentation summarized

The Institutionalist approach to climate policy repairs basic defects in the social fabric and supplies publicly what otherwise would be supplied by a privately owned monopoly; see Table 5.2. The integration of several policy domains, of which climate is one, occurs via specific actions at lower administrative levels and via private sector actors, with a role for civil society. Welfare theory–based optimality is not operationalized at the centralized level, and is diversely filled in at the decentralized level. A strict public–private separation is maintained to avoid collusion and usurpation of political power by private interests. Private monopolies endanger relevant political processes and are to be avoided; in the climate and energy domains, this refers to natural monopolies in energy transport

Table 5.2 Instrument types appropriate for Institutionalism

1. Institutional framework	Preventing private monopolies; generic emission pricing; real-time electricity market
2. Public infrastructure, spatial planning and standardization	Public supergrid; public infrastructure for hydrogen transport and storage
3. Publicly funded R&D and Demonstration	Focus on basic R&D; ARPA-type stimulants; demonstration of long-term-high-risk technologies
4. Informational instruments	Background information for framing private decisions
5. Financial instruments	No technology-specific instruments
6. Standards and regulations	No technology-specific instruments

systems, which must strictly be avoided. Temporary monopolies based on intellectual property rights present the same danger, but comprise a subject that only tangentially touches on climate policy.

Generic emission pricing can be set up via upstream administration of a tax on fossil fuel inflow, with a refund on outflow as a sort of deposit system. No refund is given for CO_2 emitted, constituting an emission tax. The new electricity market has to deal with two peculiarities. Electricity produced must be used directly, if not functionally then by costly temporary storage in batteries, pumped hydro and heat. Its production with renewables is variable and irregular, and its final use is variable and irregular. Old market designs will have to be replaced by real-time variable markets equalizing supply and demand. This is enabled by new, fast-switch technologies, also applied at a decentralized level.

Public infrastructure links to several levels of government, with the supergrid having international linkages, as well. The supergrid connects regional high-tension grids increasingly cheaply, based on the use of new HVDC technologies. By enlarging the grid over time zones and climate zones, irregularities in supply and demand can be evened out, costs reduced and supply security increased. The piped transport grid for hydrogen is monopolistic and hence public, as is storage during large-scale development. Additional private decentralized distribution options are possible, comparable to present-day gasoline stations. R&D will shift to basic research, mostly exploratory, with ARPA-type subsidies for exceptional high-risk/high-potential subjects. Some high-risk subjects can be brought to the demonstration and implementation level publicly, as occurred with the monopolistic internet.

By creating incentives and options, it is possible to largely avoid using technology-specific instruments for emission reduction in climate policy. Only where deep lock-ins in high-emitting activities exist, as with long-lifespan capital goods, may temporary action be part of instrumentation, such as in developing zero-emission steel and aviation technologies to the point where they can be implemented on a large scale. The Institutionalist approach also allows for deep decentralization of government, whereby cities are not assigned tasks but are given the space to do what is advantageous and interesting for their citizens.

Notes

1. Attempts to broaden the WTO to involve development and equity issues broke down in the Doha round in 2015.
2. See Chapter 12 of Faure and Peeters (2011) for the authors' concluding survey.
3. See http://ec.europa.eu/environment/legal/liability/.
4. See the Nord Pool website for details: https://nordpoolspot.com.
5. For further literature references, see the footnotes in Stiglitz (2014) on p.2.
6. See https://www.onvia.com/about-us.
7. A 1,300-page set of documents; see https://ec.europa.eu/energy/en/news/commission-proposes-new-rules-consumer-centred-clean-energy-transition.

References

Acemoglu, D., Johnson, S., & Robinson, J. A. (2005). Institutions as a fundamental cause of long-run growth. In P. Aghion & S. M. Durlauf (Eds.), *Handbook of economic growth, Vol 1A*. Amsterdam: North-Holland/Elsevier.

Alvaredo, F., Chancel, L., Piketty, T., Saez, E., & Zucman, G. (2017). Global inequality dynamics: new findings from WID.World. *American Economic Review, 107*(5), 404–409. doi: 10.1257/aer.p20171095

Anthoff, D., Hepburn, C., & Tol, R. S. J. (2009). Equity weighting and the marginal damage costs of climate change. *Ecological Economics*, 68(3), 836–849. doi: 10.1016/j.ecolecon.2008.06.017

Aristotle. (330–322 BC). Constituton of Athens. (1920, F. G. Kenyon trans.). In J. Barnes (Ed.), *The complete works of Aristotle. The revised Oxford translation* (vol. 2, pp. 2341–2383). Princeton, NJ: Princeton University Press.

Backhaus, K., Gausling, P., & Hildebrand, L. (2015). Comparing the incomparable: lessons to be learned from models evaluating the feasibility of Desertec. *Energy, 82*, 905–913. doi: 10.1016/j.energy.2015.01.100

Baker III, J. A., Feldstein, M., Halstead, T., Mankiw, N. G., Paulson Jr, H. M., Shultz, G. P., ... Walton, R. (2017). *The conservative case for carbon dividends*. Retrieved from https://www.clcouncil.org/

Baumol, W. J. (1972). On taxation and control of externalities. *American Economic Review, 62*(3), 15.

Baumol, W. J., & Oates, W. E. (1971). The use of standards and prices for protection of the environment. *Swedish Journal of Economics, 73*(1), 12.

Bicer, Y., & Dincer, I. (2017). Life cycle evaluation of hydrogen and other potential fuels for aircrafts. *International Journal of Hydrogen Energy, 42*(16), 10722–10738. doi: 10.1016/j.ijhydene.2016.12.119

Bodin, J. (1992). *Bodin: on sovereignty*. Cambridge, England: Cambridge University Press. (Originally published 1576.)

Boldrin, M., & Levine, D. K. (2008). *Against intellectual monopoly*. Cambridge, England: Cambridge University Press.

Brenton, A. (2013). 'Great Powers' in climate politics. *Climate Policy, 13*(5), 541–546. doi: 10.1080/14693062.2013.774632

Broadbent, J. (2002). Japan's environmental regime: the political dynamics of change. In U. Desai (Ed.), *Environmental politics and policy in industrialized countries* (pp. 295–355). Cambridge, MA: MIT Press.

Capano, G., Rayner, J., & Zito, A. R. (2012). Governence from the bottom up: complexity and divergence in comparative perspective. *Public Administration*, 90(1), 56–73. doi: 10.1111/j.1467-9299.2011.02001.x

Crawford, R. H., Bartak, E. L., Stephan, A., & Jensen, C. A. (2016). Evaluating the life cycle energy benefits of energy efficiency regulations for buildings. *Renewable and Sustainable Energy Reviews*, 63, 435–451. doi: 10.1016/j.rser.2016.05.061

Dong, Y., Ishikawa, M., & Hagiwara, T. (2015). Economic and environmental impact analysis of carbon tariffs on Chinese exports. *Energy Economics*, 50, 80–95. doi: 10.1016/j.eneco.2015.04.008

Droege, S., Asselt, H. v., Das, K., & Mehling, M. (2016). *The trade system and climate action: ways forward under the Paris Agreement* (Making the International Trade System Work for Climate Change Working Paper, ed.). London: Climate Strategies. Retrieved from http://www.climatestrategies.org.

EU. (2012). *Energy efficiency directive 2012/27/EU of the European Parliament and of the Council*. Luxembourg: Official Journal of the European Union.

Falkner, R. (2015). International negotiations: towards minilateralism. *Nature Climate Change*, 5(9), 805–806. doi: 10.1038/nclimate2767

Falkner, R., Stephan, H., & Vogler, J. (2010). International climate policy after Copenhagen: towards a 'building blocks' approach. *Global Policy*, 1(3), 252–262. doi: 10.1111/j.1758-5899.2010.00045.x

Faure, M., & Peeters, M. (2011). *Climate change liability*. Cheltenham, England: Edward Elgar.

Feenstra, R. (2009). *Hugo Grotius mare liberum 1609–2009: original Latin text and English translation*. Leiden, Netherlands: Brill.

Grossman, G. M., & Lai, E. L.-C. (2004). International protection of intellectual property. *American Economic Review*, 94(5), 1635–1653. doi: 10.1257/0002828043052312

Heller, M. A., & Eisenberg, R. S. (1998). Can patents deter innovation? The anticommons in biomedical research. *Science*, 280(5364), 698–701.

Hobbes, T. (1968). *Leviathan: Edited by CB Macpherson*. Harmondsworth, England: Penguin Books. (Originally published 1651.)

Huppes, G. (1993). *Macro-environmental policy: principles and design*. Amsterdam: Elsevier.

Huppes, G. (2011). *Instruments for effective global climate policy: the carbon deposit system*. Leiden, Netherlands: CML, Leiden University. Downloadable. https://openaccess.leidenuniv.nl/handle/1887/18319.

Huppes, G., & Huele, R. (2018). Institutionalist climate governance for pleasant cities and the good life. In P. Spinozzi & M. Mazzanti (Eds.), *Cultures of sustainability and wellbeing: theories, histories and policies*. Abingdon, England: Routledge.

IEA. (2016) *World Energy Outlook 2016*. Paris: International Energy Agency.

Jaffe, A. B., & Lerner, J. (2011 (2004)). *Innovation and its discontents: how our broken patent system is endangering innovation and progress, and what to do about it*. Princeton, NJ: Princeton University Press.

Kopits, E., Marten, A., & Wolverton, A. (2013). Incorporating 'catastrophic' climate change into policy analysis. *Climate Policy*, 1–28. doi: 10.1080/14693062.2014.864947

MacKay, D. J. C., Cramton, P., Ockenfels, A., & Stoft, S. (2015). Price carbon — I will if you will. *Nature*, 526(7573), 315–316. doi: 10.1038/526315a

Martín-Barrera, G., Zamora-Ramírez, C., & González-González, J. M. (2017). Impact of flexibility in public R&D funding: how real options could avoid the crowding-out effect. *Renewable and Sustainable Energy Reviews*, 76, 813–823. doi: https://doi.org/10.1016/j.rser.2017.03.086

Mazzucato, M. (2013). Why innovation needs the help of an active state. *Financial Times*, August 21, 2013.

Metcalf, G. E. (2008). Designing a carbon tax to reduce U.S. greenhouse gas emissions. *Review of Environmental Economics and Policy*, *3*(1), 63–83. doi: 10.1093/reep/ren015

Milanovic, B. (2016). *Global inequality: a new approach for the age of globalization*: Cambridge, MA: Harvard University Press.

Min-Fin-BC. (2014). *Carbon tax. Overview of the revenue-neutral carbon tax*. Vancouver, Canada: Ministry of Finance, British Columbia. Retrieved from http://www.fin.gov.bc.ca/tbs/tp/climate/carbon_tax.htm.

Mokyr, J. (2004). *The gifts of Athena: historical origins of the knowledge economy*. Princeton, NJ: Princeton University Press.

Moser, P. (2013). Patents and Innovation: evidence from economic history. *Journal of Economic Perspectives*, *27*(1), 23–44. doi: 10.1257/jep.27.1.23

Nordhaus, W. (2015). Climate clubs: overcoming free-riding in international climate policy. *American Economic Review*, *105*(4), 1339–1370. doi: 10.1257/aer.15000001

Nussbaum, M., & Sen, A. (1993). *The quality of life*. Oxford, England: Oxford University Press.

O'Neill, E. J. (1990). Current international developments. Approaches and developments in the cleanup of waste sites in Europe. *Environmental Claims Journal*, *3*, 127–134.

Ostrom, E. (2010). Beyond markets and states: polycentric governance of complex economic systems. *American Economic Review*, *100*(3), 641–672. doi: 10.1257/aer.100.3.641

Pickett, K. E., & Wilkinson, R. G. (2015). Income inequality and health: a causal review. *Social Science & Medicine*, *128*, 316–326. doi: 10.1016/j.socscimed.2014.12.031

Pigou, A. C. (1932). *The economics of welfare*. Indianapolis, IN: Library of Economics and Liberty. (Originally published 1920.) Downloadable from http://www.econlib.org/library/NPDBooks/Pigou/pgEW.html.

Piketty, T. (2014). *Capital in the twenty-first century*. Cambridge, MA: The Belknap Press of Harvard University.

Portney, P. R., & Weyant, J. P. (1999). *Discounting and intergenerational equity*. Washington, DC: Resources for the Future.

Rayner, S. (2010). How to eat an elephant: a bottom-up approach to climate policy. *Climate Policy*, *10*(6), 615–621. doi: 10.3763/cpol.2010.0138

Revenue & Customs. (2012). *Carbon price floor*. London: HMRC. (Originally published 2010.)

Rohrabacher, D., & Crilly, P. (1994). Case for a strong patent system. *Harvard Journal of Law & Technology*, 8, 263.

Rosselló-Batle, B., Ribas, C., Moià-Pol, A., & Martínez-Moll, V. (2015). An assessment of the relationship between embodied and thermal energy demands in dwellings in a Mediterranean climate. *Energy and Buildings*, *109*, 230–244. doi: 10.1016/j.enbuild.2015.10.007

Rothe, D. (2016). Energy for the masses? Exploring the political logics behind the Desertec vision. *Journal of International Relations and Development*. *19*(3) 392–419. doi: 10.1057/jird.2014.17

Schroeder, H., Burch, S., & Rayner, S. (2013). Novel multisector networks and entrepreneurship in urban climate governance. *Environment and Planning C: Government and Policy*, *31*(5), 761–768.

Shapiro, C. (2001). Navigating the patent thicket: cross licenses, patent pools, and standard setting. In A. B. Jaffe, J. Lerner, & S. Stern (Eds.), *Innovation policy and the economy* (pp. 119–150). Cambridge, MA: MIT Press.

Stern, N. (2008). The economics of climate change. *American Economic Review*, 98(2), 1–37. doi: 10.1257/aer.98.2.1

Stiglitz, J. E. (2014). *Intellectual property rights, the pool of knowledge, and innovation.* (Working Paper No. 20014). Cambridge, MA: U.S. National Bureau of Economic Research.

Tamiotti, L. (2011). The legal interface between carbon border measures and trade rules. *Climate Policy*, *11*(5), 1202–1211. doi: 10.1080/14693062.2011.592672

Tol, R. S. (2008). The social cost of carbon: trends, outliers and catastrophes. *Economics* 2(2008-25), 1–22.

UNFCCC. (2016). *Guidance to assist developing country parties to assess the impact of the implementation of response measures, including guidance on modelling tools* (2016/4). Bonn, Germany: UNFCCC. Retrieved from http://unfccc.int/resource/docs/2016/tp/04.pdf.

Verzijlbergh, R., De Vries, L. J., & Lukszo, Z. (2014). Renewable energy sources and responsive demand. Do we need congestion management in the distribution grid? *Power Systems, IEEE Transactions*, 29(5), 2119–2128. doi: 10.1109/TPWRS.2014.2300941

Victor, D. G. (1992). Practical aspects for implementing greenhouse taxes: issues for OECD countries. In *Climate change: designing a practical tax system*. Paris: OECD.

Weitzman, M. L. (2009). On modeling and interpreting the economics of catastrophic climate change. *Review of Economics and Statistics*, *91*(1), 1–19.

Weitzman, M. L. (2011). Fat-tailed uncertainty in the economics of catastrophic climate change. *Review of Environmental Economics and Policy*, 5(2), 275–292. doi: 10.1093/reep/rer006

Xu, W., Li, Q., & Huang, M. (2015). Design and analysis of liquid hydrogen storage tank for high-altitude long-endurance remotely-operated aircraft. *International Journal of Hydrogen Energy*, 40(46), 16578–16586. doi: 10.1016/j.ijhydene.2015.09.028.

6 Climate policy instrumentation at the crossroads: prospects

The clash of political philosophies in climate policy instrumentation resolved

The distinction between central control and institutional restraint dates back to ancient political philosophy, with Plato opposing Aristotle in the Western tradition and, somewhat similarly, Confucius opposing Laozi in the Eastern tradition. Plato advocated a wise and just government having no constraints on domains of action, going to the extreme of interfering with human reproduction (Brown, 2011, Section 4.4). Aristotle, in contrast, focused on the constraints needed to bridle the reach of central authority (Aristotle, 1950). Both tendencies are always present in any society and are expressed differently per societal domain. Lack of clarity about the tendency being followed can lead to confused, hybrid and incongruent mixes in domains of application, as is the case with respect to current climate policy instrumentation.

This distinction has been a central focus of modern political philosophy since the Enlightenment. The absolutist version arose in the 16th century. Bodin advocated the absolutist state without constraints, but not yet combined with democracy. Conversely, new libertarians such as Nozick and von Mises, advocate extreme constraints, with the state safeguarding external safety only (Bodin, 1576/1992; Nozick, 1974; Rothbard, 2009). Political developments moving toward constraint of absolutist central power were most clear in the 17th century in the decentralized Dutch Republic and in England after the Glorious Revolution, heralding the Industrial Revolution; see Table 2.1 earlier. However, democracy in the aftermath of the French revolution has tended toward a centralized absolutist view in which government is seen as representative of the people, and hence knowing what is best for them. This perspective is supported by operational welfare theory. The extremes of pure absolutism and libertarianism both go one step too far, obstructing an open society. Not all societal domains can be run by government, nor should they be, while some domains, such as defense, are purely public by necessity. In industrialized societies, markets are created and regulated for private action. The domain of climate policy can be aligned with either central planning or institutionalist indirect controls. Its mixture is ill fated, similar to a partly public and partly private army.

Incongruent philosophies around the ill-fated tradable emission permits

The largest and most developed emission pricing scheme today is the EU-ETS, which consists of increasingly auctioned emission permits and covers a little over half of all EU emissions. The pricing scheme adheres to the central Planning & Control philosophy to the extent that all emitters are required to have valid emission permits. It follows that direction, as well, in that the sum total of emission permits is systematically reduced each year according to plan. It also aligns with Institutionalist philosophy by creating incentives for decentralized emission reduction. From an Institutionalist point of view the Planning & Control elements constitute serious but avoidable burdens. They make it impossible to cover all the billions of emitters, with the consequence that inequalities are created among them. Political vulnerability results, with exemptions and oversupply of permits. The planning of emissions reduction simply cannot anticipate and accommodate unavoidable private economic mechanisms. The price for permits is inherently unstable due to their fully inelastic supply, the cap, and their short-term inelastic demand. The price will either be too low to guide medium- and long-term investment, or it will be unacceptably high, curbing economic growth. Adapting the volume of permits to create a relevant and acceptable price is possible technically by adapting the volume accordingly, and then leaving the planned reduction path.

The instrument then moves toward becoming an Institutionalist version of emission pricing, but loses out from a Planning & Control perspective. Achieving only partial coverage, the administrative management problems seem quite substantial. Setting the price by regular volume adaptations seems not only a political obstacle, but is not easy to underpin empirically. What is the right price and the right volume in an economy with significant and unpredictable swings in employment and economic growth, with unpredictable technological and market developments and unstable international developments? Such uncertainties can lead to serious political problems: "not me, not now." With internationally tradable permits, moreover, national adjustments in volume are unacceptable. A country that increases the number of permits auctioned would effectively tax all other countries. As a result, price differences between national systems seem unavoidable. Stepwise movement toward the Institutionalist pricing instrument is perfectly possible: set the yearly volumes to realize the relevant price path, move the administrative implementation upstream to cover all emitters equally, and shift from delayed administrative implementation to transaction-based implementation as is done with duties and excises. This is then a clear Institutionalist solution.

The more coherent development of Planning & Control instrumentation makes the institutionalist part redundant. Specific emissions reductions are to be advanced more speedily than pure cap-and-trade approaches can induce them, additional measures must be added, as has been the case in the EU. When a country is very successful in achieving emissions reduction, the demand for emission permits declines and so does their price. Under the cap-system others can increase their emissions, more cheaply. This occurred initially in Germany after policies introducing solar and wind power were implemented. Successful Planning & Control instruments make the administratively complex cap-and-trade system superfluous. If not intensively maintained it will collapse with the spread of new loopholes, fraud and exemptions. An active exit is then the preferred route.

The current stalemate between the two governance visions around cap-and-trade policies prevents the development of a primary, efficient, Institutionalist pricing mechanism: the upstream-administered encompassing emission tax and open-to-all, large-scale, real–time priced electricity markets. Adjoining planning instruments designed to induce renewables instead fragment electricity markets, which is detrimental for the large-scale deployment of renewable energies. The stalemate also prevents the clear development of effective Planning & Control instrumentation, as long as it is assumed that the cap-and-trade system will basically do its job. But to be effective, Planning & Control instrumentation must set an overall reduction goal and realize it, not just add measures to the cap-and-trade system to speed up change. We are at a real crossroads, and a choice must be made if we want to achieve effective climate policy instrumentation.

Two governance strategies for climate policy instrumentation

Consistent pure governance approaches emerging

Starting from either of the two pure governance approaches can provide a clear path toward effective emission reduction: through sectoral planning, as further detailed in Planning & Control, and, in Institutionalism, through repairs in the institutional framework of society to create incentives and options for moving toward a zero-emission society. A first step is to develop both options for instrumentation, as done in chapters 4 and 5, and compare them from the perspective of climate change. How feasible are they, politically and administratively? How cost-effective? And what is the likelihood that they will reduce emissions close to zero, enabling us to meet the 2-degree or 1.5-degree climate goal? Ultimately, however, the choice between instrumentations will not be based on such project-type evaluations. Governance choice is related more broadly to how political philosophy will shape a society's development, with climate policy as one major domain among others.

An unavoidable choice

Evaluation of strategic governance choices involves traditional elements related to effectiveness and efficiency – in the case of climate, relating to CO_2 emission reduction – with subcategories that include political vulnerability and complexity of administrative implementation. The governance domain also has broader links to legitimacy and the lack of it, expressed in noncompliance and retraction into populism. Broader still, governance links to the psychological domain, relating to independence, creativity and to what Aristotle called "the good life." That life involves active participation in public affairs, by necessity requiring decentralized autonomy in broad spheres of public life.

A deep ideological reference to governance strategies is, for the most part, not directly linked to party politics. Central planning is preferred in some domains, if not required technically. Control of infectious diseases and of the electricity supply must be organized centrally; so, too, national defense and management of the legal system. There are good reasons to centralize and unify in virtually all domains of public policy. But there are also good reasons to leave decisions to more autonomous decentralized and private parties, who have their own circumstances, knowledge, wishes and history. If all is said and done based on instructions from above, life becomes unbearable for all but those at the top. Dictating the good of all from above may, in fact, not lead to the good of all. This tension cannot be resolved, but solutions can be offered that consider the different options. General concern about disruptive climate change can be channeled into relevant central measures – from a technical perspective, the more central the better. This concern may also be channeled into incentives to stimulate decentralized ingenuity and action. Pursuing through centralized policies what is technically best for all is a huge and daunting task in a dynamic, market-based society, essentially steered by institutions. It can easily backfire because complex, centralized, political processes are sometimes dead wrong and are easily manipulated by domain specialists. Diesel fumes appeared to be clean on paper but were toxic in practice. Especially in a mass media-oriented, internet society, special interests may have the greatest ability to influence specific items that end up on the political agenda, and thus what is decided centrally.

A choice for Planning & Control or for Institutionalism can rightfully be made in most policy domains. These options exist from left to right across the political spectrum and in most countries, including communist nations. Only extreme left-wing and right-wing libertarianism seem to exclude active climate policy in both of the two governance paths. Mainstream social, liberal and conservative political views are concerned with how people live and with the course of societal development. They agree on the risks posed by climate change, but not on the solutions. Also within specific parties there is no consensus. Technocratic solutions are always a possibility, and justifiably so for clear and specific problems. Flue gases from coal-fired power stations can be cleaned, and such scrubbing mechanisms have been highly effective in most industrialized countries, decreasing sulfur oxide and nitrogen oxide emissions by more than 80% over two decades. The choice is more complicated if the problem to be solved involves broad domains of activity by virtually all actors in society, as is the case with CO_2 emissions. An effective technocratic solution then becomes a very large and complex task. It may remain an elusive task in a situation of only limited control in a dynamic, globalizing economy. Institutions and options may not be effective in all domains, and even in the climate domain, not for emissions other than CO_2. For a given goal, the choice between Planning & Control and Institutionalism is then not only one of political philosophy, but also one of effectiveness.

Current climate policy instrumentation is at a crossroads in terms of governance direction. What should be centralized to the national level and higher and what should be decentralized to the state or regional level, to local government and to private parties and persons? From a Planning & Control perspective, the strategic entry into climate policy involves target setting, technologies specification and top-down implementation processes based on democratic, centralized sovereignty in the nation-state. However, detailed targets and regulations make local initiatives less relevant. This is one main mechanism detracting from dynamic efficiency improvement. Such actions of supporting centralization for reasons pertaining to efficiency are also opposed to decentralization of power. Institutionalism as developed here for climate policy centralizes generic aspects of society in the most abstract way possible and does not set targets. To achieve widespread decentralization, such institutions must be redesigned in terms of decentralization of government power as well as decentralization of activities to private parties such as firms, consumers and citizens and their organizations.

Political-administrative feasibility of the two instrumentations

One core feasibility question relates to the political processes required to actively reverse undesirable tendencies and create desirable ones, including those surrounding climate policy. The political spectrum can be subdivided into three main streams: socio-liberal democratic welfare optimization, the traditional conservative tendency toward social stability, and the radical anarcho-capitalist tendency toward reducing virtually all state tasks. Only the first two have a genuine connection with climate policy. The third, more fundamentalist anarcho-capitalist direction, is opposed to all climate policy, viewing the cure as worse than the disease. This perspective is present in the United States among ultra-libertarians and Tea Party Republicans. Nonetheless, proponents of this direction would certainly favor the Institutionalist approach over the thicket of regulations arising from Planning & Control. Democratic welfare theory, currently dominant, can support both Planning & Control and Institutionalism. Within the sphere of technocratic planning direction some advocates may favor aspects of the Institutionalist approach as being more efficient and effective in reducing climate change than a multitude of specific policy actions, which can hardly create dynamic mechanisms. Technical improvements in tradable emission permits might then move in an Institutionalist direction by stabilizing prices, simplifying administrative procedures and increasing the domain of application. An Institutionalist-encompassing upstream-administered emission tax could then result for efficiency reasons alone. Altogether, there may be majority support for an administratively simpler Institutionalist-type of instrumentation.

Efficiency and effectiveness

The current hybrid Planning & Control approach has not yet led to a substantial reduction of global CO_2 emissions. Atmospheric concentrations continue to rise more rapidly than ever; see Figure 1.1 earlier. Reconsideration of the current setup of climate policy seems wise from this global point of view, with either better Planning & Control or Institutionalism as the main alternatives. Effectiveness and efficiency in climate policy are usually linked to Planning & Control governance, trying to do now what is necessary, but without real overall strategic justification. Operational optimization works medium-term at best, not for creating near zero emission society, as Grubb and co-authors, convincingly state. Deep reductions require deep transformation over the long term, a mechanism that is not open to quantified optimization (Grubb, Hourcade, & Neuhoff, 2014). By creating dynamics, Institutionalism may inherently be more effective and efficient long term, as historical analysis of the role of institutions for economic growth indicates. Abolishing central target setting, which follows from Institutionalism, allows for decentrally empowered cities and regions, more power for civil society, and more empowered businesses and consumers. Long term effectiveness and efficiency considerations alone provide reasons for shifting to Institutionalist instrumentation now.

Legitimacy, decentralization and the "good life"

Legitimacy is a precondition for stable government and governance. The lack of it may lead to incoherent changes in policies and instrumentation, guided only by short-term issues and lacking the trust needed for always-complex policy integration. Legitimacy, trust and support are here taken broadly to include informal processes beyond the legal-administrative realm, including concepts such as cultural legitimacy (Kailitz, 2013, pp. 43–44) and (Greene, 2015, section 3). By involving civil society in the preparation of policy instrumentation and implementation, with a key role for the decentralized level, trust can be created in layered processes (Luhmann, 1968/1979). Legitimacy, effective climate policy and well-being are directly connected in Institutionalist climate policy instrumentation. These processes are substantially based in the public space at the city level in autonomous cities. The role of cities, or more generally urbanities, would then become very different, moving from guided climate policy implementation to decentralized autonomy with responsibility. Autonomy, responsibility and well-being are closely connected in Aristotle's good life and are closely connected to political autonomy: "[a human] "has the perception of good and bad and right and wrong and the other moral qualities, and it is in partnership with these things that makes a household and a city-state" (Aristotle, 1932/1998, book 1, section 11). The autonomy referred to pertains to the most decentralized level, to persons and their families within the city-state, as opposed to the collective good of Plato. Institutionalism pursues safeguards for decentralization by controlling public and private monopolistic tendencies, in line with Aristotle's *Constitution of Athens* (Aristotle, 330–322 BC).

Current psychological investigations relate autonomy to intrinsic motivation, self-regulation and well-being (Ryan & Deci, 2000). Thwarting autonomy and competence of the individual diminishes motivation and well-being. Empirical analysis of educational success also supports the importance of control versus autonomy. Under direct control, students progress less than when given active autonomy (Fei-Yin Ng, Kenney-Benson, & Pomerantz, 2004). A current reality is that a significant portion of the population is disgruntled and passive, open to short-term populist manipulation. Proposed solutions centered on encouraging more voting hardly contribute to autonomic competence, and certainly not when one is voting together with tens of millions of others. "We the people" can become a dangerous concept (Morris, 2000), simplifying even complex collective considerations to a "yes" or "no" answer, with the subjects for choice decided by a very small group of elites. The importance of democracy, defined as majority rule, may reside in a very different domain. Preventing usurpation of government power by means of institutional constraints and regular change of rulers may be a precondition for the lasting autonomy of cities as well as for the good life of the many. Decentralized autonomy and power may be a long-term precondition, but it is not enough to stabilize our climate.

The links between climate policy instrumentation, happiness, and the good life are quite direct. Institutionalist instrumentation links to decentralized autonomy within generic surroundings of incentives and options creation for deep emission reduction. Decentralized autonomy links to personal motivation and development, to personal and organizational competence: to taking one's life in one's own hands. Happiness and well-being may then follow, dependent, or course, on other conditions as well. Happiness is also a matter of circumstances. Supportive circumstances, Aristotle emphasized, are necessary and these can be difficult to organize directly. Effective climate policy certainly reduces the likelihood of climate disasters; that is, of unlucky circumstances. Promoting autonomy in politically autonomous cities and regions is a key to effective climate policy in Institutionalism. Such autonomy reduces the chances of other disasters occurring, as well: those of a political origin, with the social and economic disasters that can follow. The centralizing tendencies of Planning & Control, conversely, may contribute to dangerous tendencies, encouraging populism and eventual chaos. More direct, piecemeal issues-based voting can deliver a blow to effective climate policy, and to other policies as well.

Pure governance approaches combined?

Can climate policy instrumentation combine elements from the centralized and decentralized approaches and then avoid the pitfalls of the current hybrid solution? Institutionalism has two basic elements. One element is an encompassing upstream-administered emission tax, already partially implemented in some countries and being actively considered in others. The other element is a large-scale, real-time, equal-for-all priced electricity market, with added HVDC-connected transmission grids such as are already developing in China, northwestern Europe and parts of the United States. That market design has clear advantages for accommodating a large share of intermittent, non-dispatchable renewable power. Politically this seems a viable option, considering the diverse political backgrounds of the countries that are already involved in supergrid development. So, which planning instruments is it possible to add without jeopardizing the functioning of these two main instruments? Negatively formulated, both the differentiation of electricity prices and the fragmentation of electricity markets must be avoided, which seems connected to key elements of Planning & Control instrumentation.

What remains from the Planning & Control domain is, first, a broad range of innovation policies. These are broader in scope than R&D and may include demonstration and supported market introduction for overcoming historical lock-ins and for passing saddle points that are preventing society from moving to more optimal solutions. A higher-level emission tax could induce the same processes, but with less predictability of outcomes, and possibly slower outcomes – and also with a smaller role for centralized views on what should be developed. A broad range of ARPA-type projects is acceptable in both governance views, with unpredicted outcomes, such as with the development of the internet. Secondly, the speed of emission reduction may also be increased by actively closing down old, high-emitting installations, where sunk costs allow for cheap high-emission options to remain on the market. The instrument types involved are then temporary in nature until the old has gone. Installations with a long lifespans, where sunk costs may be substantial, are prime targets: high-emitting power stations, ships and airplanes. Dynamically lowered emission standards and subsidies for scrapping are main options. Such combinations with primarily Institutionalist instruments could be interesting politically. Subsidies and prescriptions for non-fossil energy will tend to disturb adequate market development, as is now the case with subsidized energy, from nuclear to solar and wind, and remain at variance with Institutionalism.

Climate governance in developing countries

Effective climate policy presumably depends on capable administration in surroundings conducive to clear decision making. Most countries may satisfy these requirements in principle, and certainly most do so in terms of economic development. Badly organized countries remain poor, will not produce significant emissions, and will not be capable of reducing these. Their lack of development is a serious problem but not a climate problem. A capable administration will still have a relatively limited capacity in newly developing countries. It will need a tax system to fund highly important infrastructure investments. Providing subsidies, including tax reductions, is much more of a burden in developing countries than in developed countries. Subsidies are a heavy burden in terms of political control even in developed countries, and easily subject to fraud. Basing tax proceeds on a simple, upstream-administered emission tax seems attractive, thereby tending toward Institutionalist instrumentation for such practical reasons. Electricity infrastructure and market development occur for general reasons as well, because modern industrializing society is based on a universal, reliable electricity supply. In developing countries, this is necessary for firms as well as consumers, if only to connect to the internet. More local systems could emerge under Institutionalism if inexpensive storage capacity became available. Under centrally planned governance this would be less likely as large firms retain close ties to government, enabling them to gain control of electricity infrastructure and markets. Overall, the tension between centralized policy and decentralized incentives is very similar in developed and developing countries. However, the capacity for Planning & Control is more limited in developing countries, both politically and administrative-technically.

Political prospects

The political will to develop effective climate policy has been assumed in this book, in line with the Paris Agreement. But will implementation at the national level indeed follow? The current global tendency toward one-person rule and populism, and a revival of libertarianism, is opposed to climate policy. One may be optimistic about China under Xi Jinping, who has voiced a strong urge for sustainable development, also regarding climate, and to reduce local pollution, which would lead to reduced CO_2 emissions as well. But one-person rule and nationalist populism are closely related, and the global climate issue can easily disappear from view in such regimes. Similar questions arise for the Russian Federation under Vladimir Putin and the United States under Donald Trump. After a long period of extreme neoliberalism, libertarianism is now on the rise, opposing all welfare-oriented state action. In addition, poor countries that rely on fossil fuel for their income and state funding are highly unlikely to close down their production voluntarily. In all countries with substantial fossil resources, including wealthy countries, there is substantial political pressure to at least delay effective climate action. Primary fossils production has a yearly value on the order of ten trillion dollars. A choice in either direction of governance at the crossroads of climate policy is essential to overcome these obstacles.

Planning & Control requires a globalist view for action, with a global welfare orientation and with global participation led by an international organization such as the United Nations. Nationalist leaders, populist tendencies and strong fossil fuel interests clearly jeopardize the materialization of such climate policy. International free-riding, by defection and deceit, may break down the national support for collective climate action. Proponents of institutionalism have general ideas on how society, including global society, can best be developed: with central principles and decentralized action. It is unlikely that such a reflection on the internal structure of society can reach political majorities now. Self-steering of society involves tedious intellectual tasks, described by Niklas Luhmann as 'autopoiesis' (1985; 1995). But ultimately political majorities must materialize. Could an alliance be formed with centralist Planning & Control proponents, and would that be sufficient to maintain a stable majority? In the broad spectrum of political philosophies, left-wing and right-wing libertarians and ultra-neo-liberals oppose effective climate action, as do most populist factions. How might a stable majority be attained? A broader coalition could consist of groups in Planning

& Control that are focused on climate effectiveness; Institutionalists focused on incentives, option creation and decentralization; and traditional conservatives focused on social and political stability. Planning & Control proponents and conservatives value a strong state, but conservatives eschew control over the private economy. Institutionalists and conservatives value stability. Planning & Control and Institutionalism value welfare and happiness but see different routes toward this goal. Instrumentation around relevant market creation, an encompassing emission tax and significant support for private technology development could be attractive to all three political philosophies.

In terms of countries involved, those with the largest amounts of fossil fuel reserves and the strongest fossil-based interest groups will be the most reluctant to give up the value of their fossil resources and industries. These are also the countries with strong nationalist leaders, broadly present nationalist-populist views and significant libertarian influence. They want to profit from new technologies but not give up the old. They will not be likely to form the core of an effective international climate coalition. Instead, such a coalition could first form among the European Union countries and Japan – countries having only limited fossil reserves and a relatively global attitude. This Kyoto–EU or Kyoto–Paris–Bonn–London axis might begin as a mini "climate club"-- the coalition of the willing. Other countries and regions would be most welcome to join that club, and would be actively incentivized to do so.

References

Aristotle. (330–322 BC). Constituton of Athens. (1920, F. G. Kenyon trans.). In J. Barnes (Ed.), *The complete works of Aristotle. The revised Oxford translation* (vol. 2, pp. 2341–2383). Princeton, NJ: Princeton University Press.

Aristotle. (1950). *Aristotle's Constitution of Athens and related texts. Translated with Introduction and Notes by Kurt von Fritz and Ernst Kapp* (K. von Fritz & E. Kapp Eds.). New York, NY: Hafner Press, McMillan Publishing Co.

Aristotle. (1998). *Politics* (H. Rackham trans.). Cambridge, MA: Harvard University Press. (Originally published 1932.)

Bodin, J. (1992). *Bodin: on sovereignty*. Cambridge, England: Cambridge University Press. (Originally published 1576.)

Brown, E. (2011). Plato's ethics and politics in The Republic. In E. N. Zalta (Ed.), *The Stanford encyclopedia of philosophy*. Retrieved from http://plato.stanford.edu/archives/win2011/entries/plato-ethics-politics/.

Fei-Yin Ng, F., Kenney-Benson, G. A., & Pomerantz, E. M. (2004). Children's achievement moderates the effects of mothers' use of control and autonomy support. *Child Development*, *75*(3), 764–780. doi: 10.1111/j.1467-8624.2004.00705.x.

Greene, A. R. (2015). Consent and political legitimacy. In D. Sobel, P. Vallentyne, & S. Wall (Eds.), *Oxford studies in political philosophy* (vol. 2, pp. 71–97). Oxford, England: Oxford University Press.

Grubb, M., Hourcade, J. C., & Neuhoff, K. (2014). *Planetary economics: energy, climate change and the three domains of sustainable development*. Abingdon, England: Taylor and Francis.

Kailitz, S. (2013). Classifying political regimes revisited: legitimation and durability. *Democratization*, *20*(1), 39–60. doi: 10.1080/13510347.2013.738861.

Luhmann, N. (1979). *Trust and power*. Chichester, England: John Wiley and Sons. (German original 1968.)

Luhmann, N. (1995) *Social Systems*. Stanford, CA: Stanford University Press. (German original 1984.)

Luhmann, N. (1986) "The autopoiesis of social systems." 172–92 In F. Geyer and J. van der Zouwen, *Sociocybernetic Paradoxes: Observation, Control and Evolution of Self-Steering Systems*. London: Sage Publications.

Morris, C. W. (2000). The very idea of popular sovereignty: "we the people" reconsidered. *Social Philosophy and Policy*, *17*(1), 1–26. doi: 10.1017/S026505250000251X.

Nozick, R. (1974). *Anarchy, state, and utopia*. New York: Basic Books.

Rothbard, M. N. (2009). *Man, economy, and state*. Auburn, AL: Ludwig von Mises Institute.

Ryan, R. M., & Deci, E. L. (2000). Self-determination theory and the facilitation of intrinsic motivation, social development, and well-being. *American Psychologist*, 55(1), 68. doi: 10.1037/0003-066X.55.1.68.

Index

Page numbers referencing figures are followed by an "*f*"; page numbers referencing tables are followed by a "*t*."